AF566708

ADVANCES IN GEOPHYSICS

VOLUME II

Advances in GEOPHYSICS

EDITED BY

H. E. LANDSBERG
U. S. Weather Bureau
Washington, D. C.

VOLUME 2

1955
ACADEMIC PRESS, New York and London

ACADEMIC PRESS INC.
111 Fifth Avenue, New York, New York 10003

United Kingdom Edition published by
ACADEMIC PRESS INC. (LONDON) LTD.
Berkeley Square House, London W.1

Library of Congress Catalog Card Number: 52-12266

Second Printing, 1967

PRINTED IN THE UNITED STATES OF AMERICA

LIST OF CONTRIBUTORS

HUGO BENIOFF, *Department of Seismology, California Institute of Technology, Pasadena, California*

IRVING I. GRINGORTEN, *Air Force Cambridge Research Center, Cambridge, Massachusetts*

K. L. S. GUNN, *McGill University, Montreal, Canada*

WALTER HITSCHFELD, *McGill University, Montreal, Canada*

J. LAURENCE KULP, *Lamont Geological Observatory (Columbia University), Palisades, New York*

J. S. MARSHALL, *McGill University, Montreal, Canada*

WILLARD J. PIERSON, JR., *Department of Meteorology and Oceanography, College of Engineering, New York University, New York, New York*

Foreword

The first volume of Advances in Geophysics was well received by our colleagues. This encourages us to send this second volume into the world. It is again a collection of articles cutting widely across the various subfields of our science. This diversified contents is presented even though a few reviewers suggested that each of these books treat with progress in one area only. We are clinging, however, to our original aim: We want to acquaint specialists with advances in the neighboring sectors. Our hope is that some cross fertilization might bring forth some new ideas.

The topics presented in this volume demonstrate forcefully that new knowledge has been acquired most rapidly in areas where geophysics could take advantage of developments in other sciences. The impact of modern electronics on observing techniques has brought about a scientific revolution. We can fathom this from the two articles dealing with seismometry and radar weather. This last field, curiously enough, owes its existence to an undesired effect on equipment designed for a much less peaceful purpose. The great strides made in atomic isotope techniques have advanced geological timing tremendously as shown in the summary herein. This is certainly only a beginning because isotopic tracers are likely to produce answers to many questions in meteorology, hydrology, and oceanography also. Finally, we can see again, as in parts of Volume I, the role played by modern statistics in geophysical analysis, as shown in the articles on wind-blown waves and objective weather forecasting. As geophysical observations accumulate at an ever increasing rate, the new statistical keys are likely to open the secrets locked in the mass of data.

While this is being written, plans are well under way for a third volume. It will take us to the present physical frontiers of the globe in the Arctic Ice Islands and will include discussions of the long-range terrestrial and extraterrestrial effects on the atmosphere, the optics of the air and its suspensions, the structure of the earth beneath the continents, and the economically so important questions of groundwater.

H. E. Landsberg

February, 1955

Contents

Geological Chronometry by Radioactive Methods

By J. Laurence Kulp, *Lamont Geological Observatory (Columbia University), Palisades, N. Y.*

Earthquake Seismographs and Associated Instruments

By Hugo Benioff, *Department of Seismology, California Institute of Technology, Pasadena, California*

Advances in Radar Weather

J. S. MARSHALL, WALTER HITSCHFELD AND K. L. S. GUNN

McGill University, Montreal, Canada

Page

1. Introduction

Radars using a wavelength of the order of 10 cm will detect rain to ranges of 200 miles or more, and snow to somewhat shorter ranges. At 10-cm wavelength, the scattering of the radiation by the precipitation, which is the basis of this detection, does not weaken the beam appreciably, so that the detection of rain at long range is not hindered by any intervening rainfall. Shorter wavelengths permit more compact and lower-powered equipment, although attenuation effects must then be considered carefully. The optimum parameters for weather radar fall fairly close to those for detecting aircraft and shipping, and are not critical. The forms of display developed for other uses are similarly adaptable. A "plan view" of the precipitation over an area of 100,000 square miles can be swept out in seconds, or a three dimensional record in a few minutes. The resulting ability of "ordinary" radar equipment to record the distribution pattern of precipitation with good resolution, in three dimensions at the closer ranges, and in plan out to the limit imposed by the earth's curvature, is the strong and simple basis of the "radar meteorology" that has developed during the past decade. This new tool for observing precipitation has come into use in the same decade that has seen similar strides in the applied science of cloud seeding, and impressive developments in the fine art of flying through storm clouds. Altogether, these advances have enormously accelerated the study of the precipitation processes. The life expectancy of hypotheses concerning these processes has been greatly abbreviated.

Radar meteorology has not been limited to the study of pattern. The intensity of the precipitation can be deduced from the strength of the received signal, and notable advances have been made in the practical application of this correlation. Radar is recommended as a simple and inexpensive rain gage. From rain-intensity gradients or from the rate of mixing of the target elements, as indicated by signal fluctuations, some notion is obtained of the intensity of turbulence within the precipitating cloud. Since water clouds have only about one-millionth the reflectivity of precipitation, they are not seen by "ordinary" radars. Yet shorter wavelengths have been used in the development of cloud radars which at short range see most types of cloud, missing some that the eye sees, but seeing some that the eye misses. These extensions of radar weather technique are hard work, although rewarding. The strong basis of the subject remains the finding—fortunate to us in this field but unfortunate to original intention—that microwave radar sees rain without even trying.

Although it is not known who was first to decide that certain radar echoes were of meteorological origin, it is certain that some operational use was made of them as early as 1942 [1]. The basic theory of these weather echoes was worked out in short order by Ryde [2], in England, on the basis of Rayleigh's [3] and Mie's [4] theory of scattering. Further theoretical developments by L. Goldstein (published in the Summary Technical Report of the Committee on Propagation, National Defence Research Committee [5]) and by several workers at the Radiation Laboratory of the Massachusetts Institute of Technology (including the work on fluctuations [6, 7]) followed. An excellent summary of this early work may be found in Kerr's [8] compilation and in other volumes of the M.I.T. Radiation Laboratory Series. Maynard [9] was probably the first to publish radar pictures in a series of excellent plan (PPI) views of storms obtained by the U. S. Navy. Bent [10] followed quickly with pictures of precipitation echoes obtained in the days of the M.I.T. Radiation Laboratory.

This early work was done during World War II, generally under military auspices. Only a fraction of it has since been published. For this reason, the dates of the publications do not necessarily reflect the time at which the work was done, nor is the early history of radar weather easily accessible. There have been good reviews, however. Kerr [8] has already been mentioned. A more up-to-date account of radar storm detection by Ligda [11], complemented by an essay on the theoretical background material by Wexler [12] are contained in the *Compendium of Meteorology*. Van Bladel [13] in Belgium is now producing a monograph in the French language, a substantial part of which is devoted to radar weather.

In spite of the great development of radar weather, the number of scientists and agencies involved in this field has remained relatively small. On the whole, there is excellent personal contact between all interested parties, and the exchange of information is probably closer than is indicated by the published literature. In furthering personal contacts, as well as in the general progress, four conferences on radar weather, at which a majority of the specialists were represented, have played a great role. The proceedings [14–16] of three of these meetings have been invaluable source material in all research and development work; and they have been a priceless help in the preparation of the present review (quite beyond the extent revealed in the references, because wherever possible, reference was made to publication in a scientific journal rather than to the less widely distributed proceedings).

Although the names of many individuals will be mentioned later on, it is probably not unfair to other agencies to single out the United States Air Force (through the Geophysics Research Directorate and the Air Weather Service) and the Signal Corps for special mention as early and continuing sponsors of a great deal of research and development work. In very recent years, interest in radar weather has, however, spread to other groups. Following the early lead of industries on the Gulf of Mexico, power companies, air lines, agricultural institutions, and regional groups have started programs for weather warning and hydrological observations by radar in a manner best suited to their own purposes.

The present review is written partly in answer to the interest attending this greatly expanding application of radar to meteorology. The aim of the writers has been to summarize the main developments, with particular emphasis on the most recent work that has come to their notice. It is hoped that perusal of this review and frequent references to the bibliography will help the reader towards a fair appraisal of advances in radar weather.

2. Precipitation Patterns

2.1. Radar Displays

The height of the precipitation pattern in the atmosphere may be anything from two to ten miles; straight-line propagation limits the range therefore to a few hundred miles. The methods available for examining the precipitation pattern in the volume accessible to a radar may best be discussed with reference to the CPS-9 radar, which has been specially designed for weather observations [17]. The beam projected by a CPS-9 weather radar is one degree in width. This beam is obtained by using a wavelength of 3.2 cm and a parabolic mirror having a width

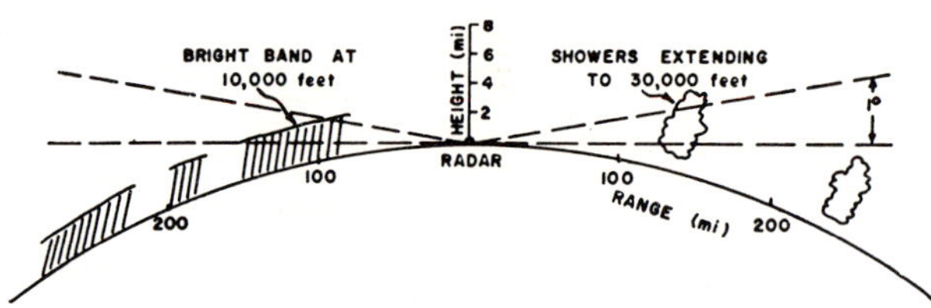

Fig. 1. Schematic diagram showing the range limitation on a weather radar due to the curvature of the earth. The vertical scale is expanded here, as it often is on vertical section displays, in order to show the precipitation masses in better detail.

of aperture of 8 ft, the beam width being approximately the ratio of wavelength to mirror diameter. Narrower beams are available, and it may be desirable, particularly when scanning in a vertical plane, to have the beam narrower in that plane and wider normal to it.

The scanning most convenient for storm detection, and amenable to operation at great range, is to set the bottom of the beam close to the horizontal, then to rotate the beam about a vertical axis (Fig. 1). The information is then displayed on a Plan Position Indicator (PPI). The earth's surface drops away from the horizontal at the radar as the square of the distance. Allowing for normal bending of the rays in the troposphere, the earth's surface is roughly 1 mile below the horizontal at range 100 miles, 4 miles at 200, and so forth. Recognizing that the siting of the equipment may prevent setting the edge of the beam quite as low as the horizontal, it can be seen that the operational range on five-mile-high thunderheads will tend to fall below rather than above 200 miles.

At shorter ranges, not likely to exceed 100 miles, the resolution in the vertical is good enough that useful vertical sections through the precipitation pattern may be obtained (Fig. 2), and displayed on a Range Height Indicator (RHI). If the equipment was designed for use with aircraft targets, it is likely to have a maximum height of scan closer to

30 degrees than 90 degrees, so that a large conical region above the radar is left unscanned. Its display is also likely to have an exaggerated vertical scale, as do Figs. 1 and 2. It is important not to carry over into one's mental picture of the precipitation pattern the distortion in shape introduced by this vertical stretching. Showers are not tall and slim, as a rule; they just tend to look that way in radar pictures.

There are other reasons for taking care in the interpretation of these pictures in vertical section. The winds aloft seldom have the same direction at all heights. The precipitation pattern, similarly, and largely due to the influence of the winds, is seldom a pattern in a plane. Further,

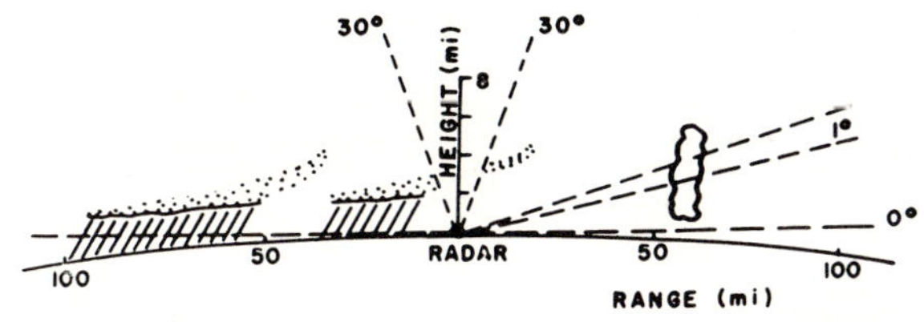

FIG. 2. A schematic vertical section similar to Fig. 1 but extending to a shorter range. It shows the vertical distortion of a typical RHI display for an antenna which scans to an elevation of 30 degrees.

rapid changes in pattern in a particular vertical plane are often attributable to the motion of the pattern-in-space across the plane of scanning. It would seem desirable, then, to study the three-dimensional pattern in three dimensions. This calls for rapid scanning throughout a hemisphere. A scanning program can be designed to do this in a few minutes. There remains then the problem of displaying the three-dimensional information. Solutions to this problem are likely to be elaborate, and so in the work reviewed here, and in work now in progress, the observations are nearly all in the form of two-dimensional patterns, either height/range, or plan, or height/time. This latter display is common in cloud observations and is very useful in studying the pattern of falling snow. It is obtained by recording the target material passing through a fixed vertically pointing beam.

A typical PPI display is shown in Fig. 3. This display is a map of the precipitation in the area surrounding the radar. The set is at the center of the picture; the concentric circles are range markers, 20 miles apart, and north is at the top of the display. The display is built up by the rotation, once every 20 sec, of a radial trace. The trace is formed by the spot (where the electron beam strikes the screen of the cathode ray tube) moving radially out from the center of the display. A new trace is started simultaneously with each pulse transmitted by the radar. The PPI trace is intensity-modulated, as it is also on the height/range and height/time displays. This means that the radar signal returned from the target is made to brighten the sweeping spot at the point on the trace corresponding to the range of the target. Intensity-modulated displays in this way map out the structure of the precipitation, and hence are used to study pattern. For measurements of radar signal strength and studies of signal

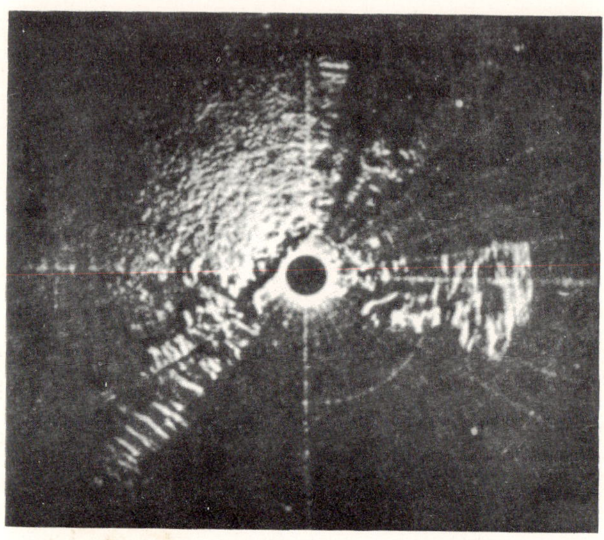

Fig. 3. A plan (PPI) view of the echoes associated with a cold front passage, obtained by holding the beam just above the horizon and rotating it about a vertical axis. (From [65].)

characteristics, an A-scope display is used. It consists of an amplitude-modulated trace, so called because the spot is displaced at right angles to its traverse by the returning signal, the size of the displacement being a measure of the intensity of the signal.

This particular PPI picture (Fig. 3) shows a typical cold front, with a line of shower activity extending from southwest to northeast. Successive pictures show the line to be moving toward the east. In advance of the frontal line of showers, there is a cluster of intense showers (discrete blobby echoes), and behind the line is a fairly large area of continuous rain (extensive pebbly echo). In vertical section, these two types of rain are also readily identifiable from their different characteristic echoes. A typical vertical section through continuous rain, as would be seen looking along a northwest bearing in Fig. 3, is shown in Fig. 4. The outstanding echo feature is the horizontal bright line of echo, the bright band. A typical vertical section through showery rain (Fig. 5) shows no horizontal bright band. The echoes tend to be more uniform in the vertical; individual echoes, usually less than 10 miles in extent, can be distinguished, and the echo outline is often irregular due to the updrafts and downdrafts associated with showers. All the echoes in these pictures are from precipitation. It is only by going to radars of higher power or shorter wavelength that the much smaller cloud particles can be detected.

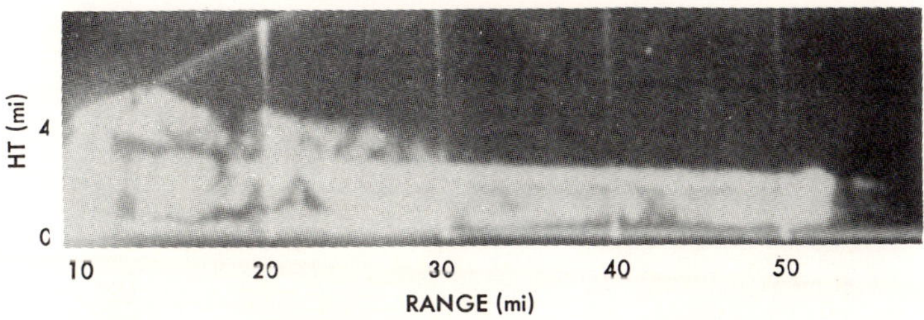

FIG. 4. Example of an RHI display showing a vertical section through continuous rain. (Photograph, Stormy Weather Group, McGill University.)

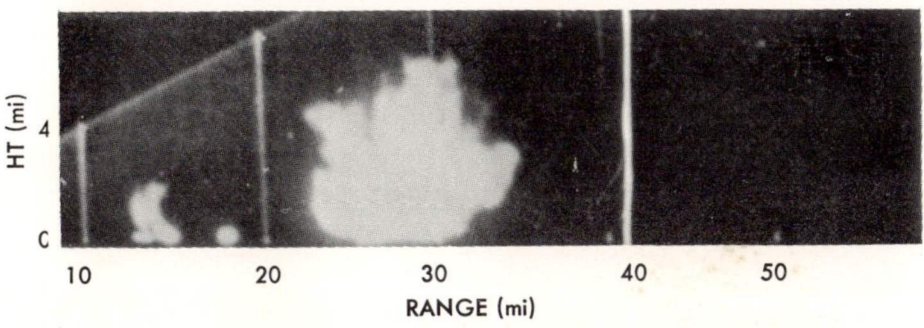

FIG. 5. Example of an RHI display showing a large shower echo extending up to about 38,000 ft. These displays (Figs. 4 and 5) avoid distortion by having nearly equal horizontal and vertical scales. (Photograph, Stormy Weather Group, McGill University.)

The radar study of precipitation patterns tends to be divided according to the two general types of precipitation observed, continuous and showery. We shall discuss patterns under these two headings.

2.2. *Continuous Precipitation*

Continuous rain as seen on a PPI (as for instance in the northwest quadrant of Fig. 3) is rather lacking in pattern, apart from the characteristic graining due to statistical fluctuations in the radar return. When this graining is reduced by averaging more independent signals, the pattern is more readily observed, but it tends to be a gradual pattern, of half-tone character, as compared with the sharp black-and-white of shower patterns. Observed in vertical section (Fig. 4), the pattern is amenable to this same description from the ground up to the intense bright band observed at or about the 0° C isotherm. Frequently vertical sections showing rain below the bright band reveal little or no target material above the band. It therefore seemed reasonable in 1947 to consider the possibility that the primary formation of this type of precipitation occurred at the 0° C isotherm [18]. A complete quantitative account of processes in the bright band still remains to be given; it is probable that several interrelated processes are involved. It does appear certain, though, that the bright band denotes the melting level of snow formed at considerably greater altitude, and that the rain falling out of this level is of the same order of intensity as the snow falling in, corresponding to the suggestion made by Ryde [2] in 1946. The study of the initiation of continuous rain is therefore essentially the study of the formation of snow.

The vertical sections through continuous rain that reveal little of the snow falling into the melting level tend to be those in which the scanning extends from the horizontal to an elevation of twenty or thirty degrees. With this type of scanning the snow only comes into the picture at considerable range. Snow has less reflectivity than rain and much less than the bright band. Further, the radiation going to it and returning to the radar traverses the bright band obliquely, and attenuation within that band is worse than that in rain. For this reason, Swingle [19], at Harvard, pointed his beam of 10-cm radiation directly upward, so that every part of the snow pattern came into the beam at a range of 5 miles or less. He recorded the returns photographically on a chart of height against time. Integration of the signals by the photographic recording, aided by the fact that no scanning was involved, gave added sensitivity to the technique and minimized the pebbling of the picture by statistical fluctuation of the signals. His pictures of the snow above the bright band resembled pictures of snow reaching the ground, obtained by scanning techniques. They revealed considerable pattern in the snow above the bright band, leading to his conclusion in 1950 that "with few exceptions, continuous surface precipitation is the result of nonuniform precipitation aloft."

The nature of snow patterns was summarized by one of the present writers [20] in 1951: pictures of snow were generally lacking in pattern. However, one characteristic pattern that did appear was a sort of "mare's tail," closely resembling the cloud formation described by that name. The precipitation appeared to develop in a short vertical element and to fall obliquely through the wind shear. The extent of the obliquity was not fully apparent until equal height and range scales were used. The resulting, almost-horizontal band might almost be mistaken for the melting band found at the 0° C isotherm.

In the same year, Bowen [21] published pictures of continuous rain and the snow above, in vertical section, obtained by scanning through the whole vertical plane. Without benefit of the photographic integration involved in Swingle's technique, Bowen's pictures were quite grainy and included only a limited region of the snow pattern. They did show upper bright bands, horizontal or nearly so, that descended at a rate rather greater than the terminal velocity of snowflakes, and brightened as they fell into the usual bright band.

Bowen's account of upper bright bands accelerated the consideration of snow patterns. Subsequently to his paper, Browne [22], Lhermitte [23], and Marshall [24] independently interpreted these bands as the

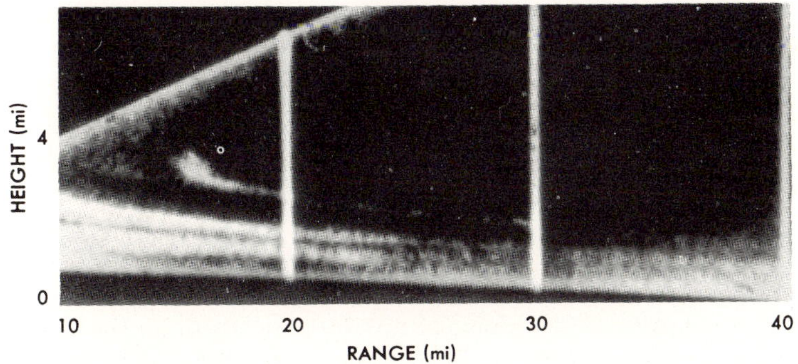

FIG. 6. The RHI display of a 3-cm radar located at Montreal Airport, 2 November 1951, showing a generating element at a height of about 3 miles, with a snow trail extending out to 40 miles range. Notice the lower, almost-horizontal trails which belong, respectively, to generating cells some 20 and 40 miles ahead (to the left) of the one shown. (From [24].)

almost-horizontal precipitation streaks that form part of the mare's-tail pattern. They formulated the theory of precipitation trails, and provided fresh observational evidence concerning these patterns, such as the example shown in Fig. 6. In this case the wind was in much the same direction everywhere, although increasing linearly with height. A more normal situation in which the trails apparently curve in three dimensions as they descend is shown in the plan view of Fig. 7.

To produce a long streak or trail, it is necessary that the snow be formed in a relatively compact generating element, and that this element continue to function for the order of an hour or more. Thus these generating elements, although disceret in space, do appear to be continuous in time. The theory of precipitation trails and trajectories is continuing to prove useful in deciphering the mechanism of snow formation and growth. The following account of it is taken from reference [24].

Let x represent distance in some horizontal direction, and z the vertical distance (as depth, increasing downward). Assume that, for a precipitation particle, $dx/dt = w(z)$, where w is the component of the wind in the x direction, and let the rate of descent of the precipitation particle be a function of height or depth, i.e., $dz/dt = v(z)$. Then for

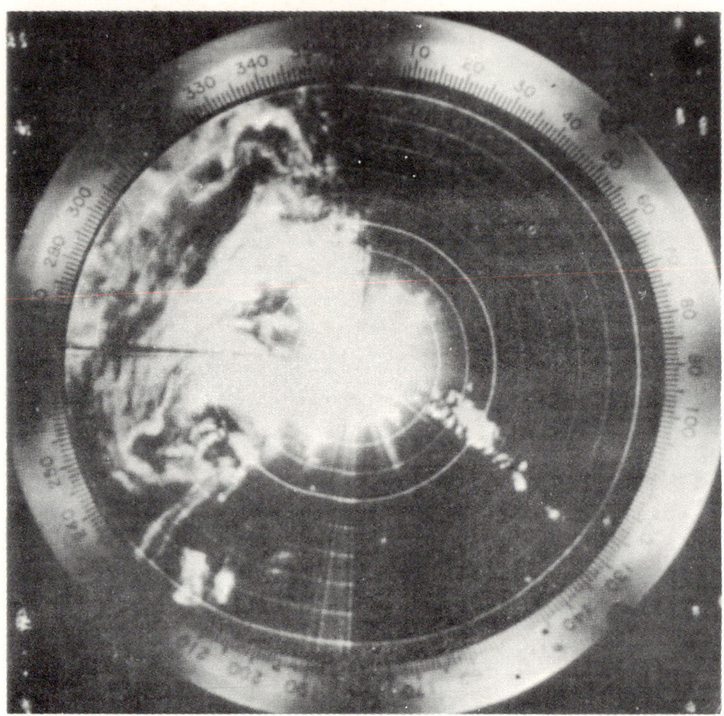

FIG. 7. The 50-mile-radius PPI display of a 3-cm AN/CPS-9 radar, showing the pattern on a day of drizzle at the surface and snow aloft. The snake-like echoes are probably due to the curvature in plan of the snow trails. (Contributed by Weather Radar Research, MIT.)

the trajectory of the particle $dx/dz = w(z)/v(z)$, and the equation of the trajectory can be stated as

$$x - x_0 = \int_0^z [w(z)/v(z)]\, dz$$

Now suppose that a small precipitation-forming element (to be treated as a point) moves with the wind at depth $z = 0$. The pattern formed by its trail of precipitation (under the assumption of a single velocity of fall) is the trajectory of the particles formed in the element, if the trajectory is referred to axes moving with the generating element. Referred to other axes, say fixed to the ground, pattern and trajectory become two different things. With $w = az$ and $v = b$, the equation becomes

$$x - x_0 = (a/2b)z^2$$

and the pattern is a parabola. The pattern will itself move as a whole horizontally, relative, say, to the ground, with velocity W, the wind at the level of formation relative to the ground.

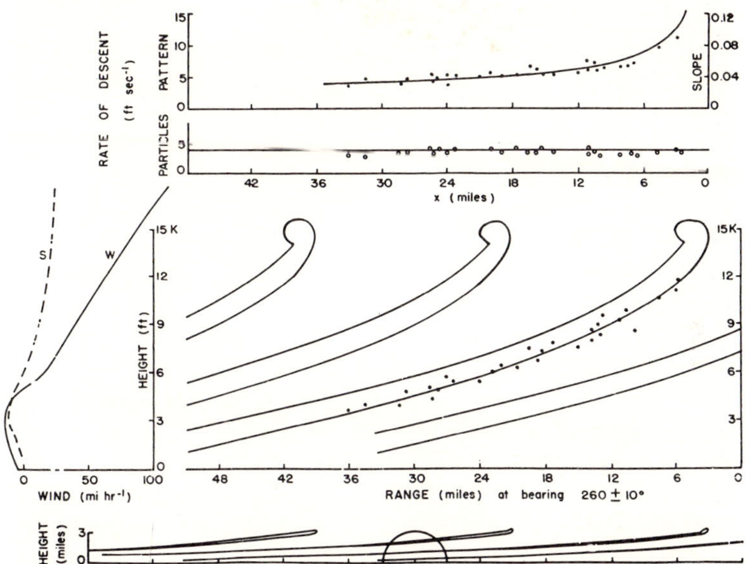

FIG. 8. Height/range data from six trails such as shown in Fig. 6 are plotted in main diagram, all with reference to a single generating element, and fitted by a parabola. These are combined with winds to give rate of descent of particles (open circles). Measured slopes of trails at various horizontal distances back from generating elements are shown at very top, with alternative scales of rate-of-descent of pattern. At bottom of figure, trails are shown against equal vertical and horizontal scales. (From [24].)

The slope of the pattern, $dz/dx = v/w$, is the vertical speed of the precipitation as a fraction of w, the difference between the wind at the depth at which the pattern is being observed and the wind at the level of formation. The rate of descent of the pattern over a fixed point on the ground, due to the pattern's movement as a whole relative to the ground with velocity W, is $W(dz/dx) = vW/w$. Thus, there is a proportionality between slope and rate of descent of the pattern, so that extremely rapid descent would not be expected to coincide with very close approaches to zero slope.

The case of 2 November 1951 is shown in Fig. 6 and analyzed in Fig. 8.

The wind increased linearly with height to 90 mi/hr at the generating level; the pattern data plotted against axes of range and height were obtained from the radar pictures, and are fitted here by a parabola. These are accumulated data from six individual patterns, some at bearing 260 degrees, some on the opposite bearing of 80 ± 10 degrees. Upon combination of the pattern data with the wind information, the rate of descent of the precipitation particles was calculated as 4 ft/sec at all heights, as shown by the open circles.

The slope of the pattern was measured directly on the radar pictures. These data, too, fit the simple theoretical curve quite well, as can be seen in the upper portion of Fig. 8. Rate of descent of the pattern over a fixed point on the earth is proportional to the slope, and is given by an alternate scale of ordinates. The slope was difficult to measure; it tended to be less than the distortion in the radar pictures, which had therefore to be calibrated carefully.

The slope is seen to decrease rapidly, with increasing distance from the generating element, to about one in twenty. Just how nearly horizontal the trails are throughout most of their length may be seen when the patterns are drawn to equal vertical and horizontal scales, as in the bottom portion of Fig. 8. If a radar on the ground were limited to a very short range, say 3 mi, as indicated by the semicircle, it would detect a succession of almost-horizontal bands descending at from 5 to 8 ft/sec, or rather more rapidly than the individual particles. If the radar scanned in a vertical plane at right angles to the wind, the slope of the band would become zero.

To help in the interpretation of such observations, Langleben [25] has recently reported measurements of the terminal velocities of aggregate snowflakes. He finds that the velocity is approximately proportional to the one-tenth power of the mass. The constant of proportionality is dependent on the basic crystal types in the aggregates and is greatly increased by riming or slight melting of the snowflakes. Earlier data on the terminal velocities of single crystals fall reasonably close to Langleben's curves. The terminal velocity determined above from a radar pattern is too high for single crystals. This, with similar evidence from pictures taken on other occasions, cited by Langleben, suggests that aggregation takes place in the generating elements, even though these occur at temperatures of $-16°$ C and below, temperatures at which aggregation would not have been anticipated. It would not be fair to suggest that only aggregates are produced in these generating elements, for the radar is prejudiced in favor of large particles. If both aggregates and single particles were present, the radar might reveal the former and not the latter.

Gunn, Langleben, Dennis, and Power [26] analyzed vertical section records of snow patterns obtained during the winter of 1951–52. They found that a specific height could be identified as the level of generation and this height was compared with heights of possible significance found from the upper air analysis. On 19 out of 22 days, some sort of trail pattern was observed; on the remaining 3 days, very little signal or pattern was observed. On the majority of days with pattern, the level of the generating elements was found to be close to a frontal surface (the boundary between two air masses aloft), and the trails were well-defined and readily identified. The top of the cloud deck associated with the frontal surface was sought in the analysis of the radiosonde data. The relative humidity was found to drop off sharply about 1300 ft, on the average, above a frontal surface, about 400 ft below the average echo top height. This study thus brings out the idea of a relatively shallow "active layer" of cloud straddling a frontal surface in which the snow generating elements develop. The days of well-defined snow trails proved to be those on which there was no instability at any height. For the rest of the 19 days with pattern, the air above the frontal surface was unstable. The patterns then consisted of only parts of trails and were generally confused. It seemed as though in the group of cases studied the instability had served only to modify the precipitation mechanism and its pattern, rather than to initiate it.

This evidence that the generating elements are associated with frontal surfaces comes from a study covering the major portion of one winter's *snowfall*. The study covered a relatively small fraction of the *number of occasions* during that winter on which snow fell. It gave no evidence of instability actually leading to the initiation of air-mass snow showers, although this process surely happens on occasion. It may be that the statistics for this group of cases in which the average amount of snowfall was great will differ materially from another grouping in which cases of light snowfall tend to be emphasized. Such a grouping is provided when records are obtained from the more sensitive zenith-pointing systems of the type described earlier in this section in connection with Swingle's work [19]. Figure 9 shows a height/time record obtained recently at McGill University with a 3-cm vertically pointing radar. Photographic recording provides signal integrating which, along with the shorter ranges of overhead targets, increases the sensitivity of the equipment. The vertical echoes at the generating level, the bright band, and the steeper slope of the rain echoes below the band show up clearly here.

Going to shorter wavelengths, still greater sensitivity is achieved and even lighter precipitation, and cloud, are detected. Although more information is provided by this increased sensitivity, interpretation may be more difficult. Atlas [27] has used for some years a vertically pointing 1.25-cm radar for cloud and precipitation observations. A sample height/time record obtained in February 1953 is shown in Fig. 10. Even though the precipitation at the ground was extremely light, about 0.05 mm/hr, the extra sensitivity gives much more detail than the displays at longer

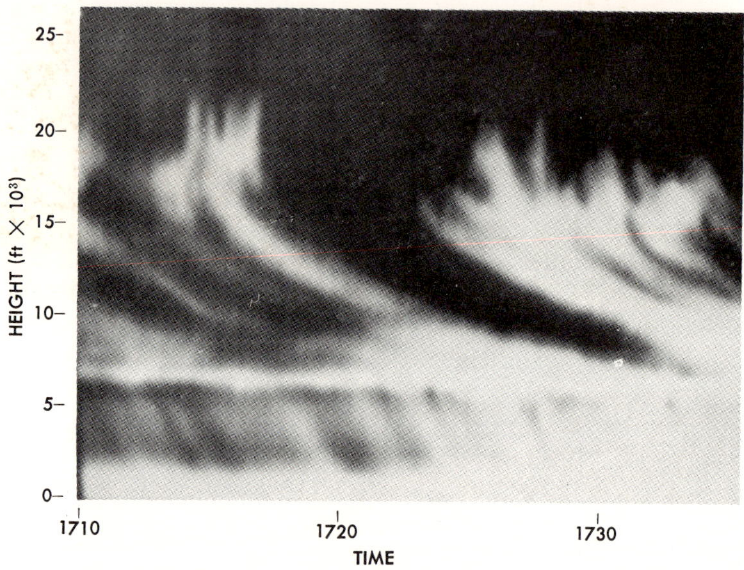

FIG. 9. The height/time display of a vertically pointing 3-cm radar located at McGill University. In this picture, recorded on 25 November 1953, the bright band, the vertical snow echoes at the generating level, the trails which become nearly horizontal at the melting level, and the steeper slope of the rain echoes below the bright band, all show clearly. The increased sensitivity obtained with the fixed vertical beam is evident from the greater detail in the snow pattern than is revealed by a height/range radar (Fig. 6). (Photograph, Stormy Weather Group, McGill University.)

wavelengths. The echoes in the top part of the figure were recorded on a moving paper, facsimile recorder. In the lower part of the figure reflectivity contours (in decibels below some reference level) are shown for a portion of the record above. These have been drawn from signal strength values obtained from a pulse integrator, which sweeps in height every minute or two. The bright band shows up as a near-horizontal line close to 2000 ft, of reflectivity 100 ± 1 db below the reference level. Cores of still higher reflectivity in the snow at 8 to 10,000 ft indicate regions of high particle density, since ice reflects only one-fifth as well as water. A 3-db difference in reflectivity represents a factor of 1.6 difference in precipitation rate. The precipitation rate in these regions of high reflectivity is thus about 0.5 mm/hr (as liquid), or about 5 mm/hr snowfall rate. The lack of echo and contours at the 4000-ft level is probably due to evaporation.

In most of the studies reported above, it has been assumed that the generating elements move with the speed of their environment. This

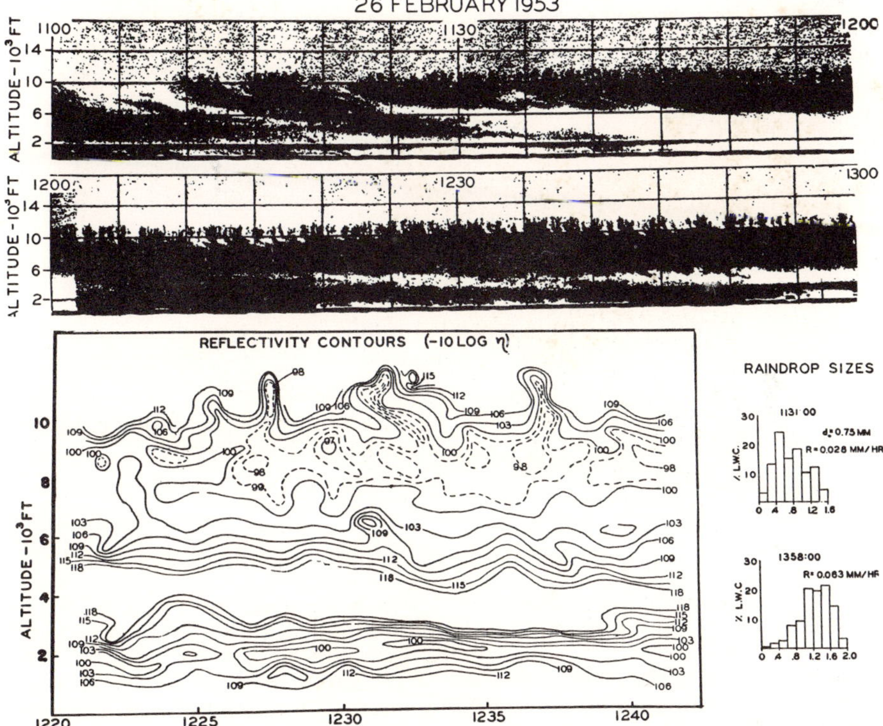

FIG. 10. The height/time display of a 1.25-cm vertically pointing radar showing echoes from light precipitation and cloud with tops at about 12,000 ft. In the lower part of the figure, reflectivity contours in decibels below some reference level have been drawn for a portion of the record above. (Courtesy, Geophysics Research Directorate, Air Force Cambridge Research Center.)

appears to be more or less the case in the observations so far analyzed. Atlas, however, in studying the records of his more sensitive equipment, has frequently noted the occurrence of "cusps," which seem to require for their explanation generating elements moving with an appreciable velocity relative to their environment.

In this discussion of snow patterns, it should be borne in mind that pattern is sometimes absent, or at any rate nearly so, as in 3 days out of the 22 studied by Gunn *et al.* Wexler and Austin [28] wisely tended to select such cases in making measurements of target strength or reflectivity against height. They analyzed these relationships to obtain the rate of growth of the falling snow. This sort of quantitative interpretation of the intensity of the radar signal requires a knowledge of the properties of the target material, particularly of its size distribution and any progressive change in that size distribution. In a way, therefore, radar appears to be less profitably employed in this form of pursuit than it is in the study of pattern. Taking a longer view, however, the short-coming is more apparent than real. The size distribution data, shape and aggregation characteristics that must be known to interpret the radar signal strength are themselves directly relevant to the physical processes of the precipitation mechanism, and the radar studies have been invaluable in stimulating their consideration.

The bright band is the outstanding feature of the continuous-rain pattern. It is very simple as an item of pattern, and a qualitative explanation of it is almost as simple. The very slight wetting of an ice particle increases its reflectivity almost to the extent that would be obtained if the particle were all water [29, 30]. Further melting cannot lead to much further enhancement, then, and may lead to a lessening of the reflectivity of the particle by bringing it closer to sphericity or by leading to the breaking up of the particle. The very slight wetting, which increases the reflectivity greatly, increases the terminal velocity only slightly [25], but the complete melting increases the terminal velocity by a factor of five or more. The number-density of particles in space is inversely proportional to their terminal velocity, and so, therefore, is the reflectivity per unit volume. Thus the achievement of a bright band is obvious as the precipitation proceeds from dry snow (poor reflectivity, slow fall) to wet snow (very good reflectivity, almost-as-slow fall) to rain (good reflectivity, fast fall).

A good quantitative account of the bright band is bound to be difficult; notable quantitative studies to date are those by Austin and Bemis [31] and Labrum [32]. As the snow approaches the melting level from above, it acquires a damp surface. This will tend to increase its reflectivity, its rate of fall, its rate of aggregation. The extent of these increases can only be estimated. As melting proceeds, the situation becomes more complicated. A small raindrop falls faster than a large snowflake, yet the smaller precipitation particles become raindrops first. There may well then be a complete reversal of relative velocities, with all the added possibilities of coalescence thereby engendered. Comparison of size distributions for rain and snow suggest that there is some breakup with melting. No precise information is available on any of these processes. The writings to date on the bright band contain valuable suggestions, but in view of these uncertainties the conclusions reached can only be regarded as tentative.

In addition to those already mentioned, numerous other workers have dealt with the bright band. The first assessment of the effect of melting, and of increased terminal velocity, on the radar signal was made by Ryde [2] in order to account for the echo enhancement at the 0° C isotherm. Signal strengths in the bright band were measured by Hooper and Kippax [33], who found them to be 5 to 9 times as great as in the rain below, in agreement with Ryde's predictions. Extensive aircraft flights above, in, and below bright-band echoes, reported by Jones [34], confirmed the presence of snow, wet snow, and rain, respectively, in these regions. Atlas and Banks [35] recognized that attenuation in the wet snow of the bright band might be considerable and pointed out how this attenuation could lead to distortion and misinterpretation of the picture of the snow region above the band. Wexler and Honig [36] have pointed out that air may be cooled considerably by melting snow falling through it. They indicate a possible operational use of radar in forecasting a change of precipitation at the ground from rain to snow, from observations of the lowering of the bright band.

The integration of ground radar and airplane flight observations has been carried out with notable success at the Massachusetts Institute of Technology. A comprehensive account of the complete precipitation region of a cyclone using radar and flight information has been given by Cunningham [37].

2.3. Showery Precipitation

In the study of showers as in the case of continuous precipitation the vertical section has proved extremely useful. An example of such an RHI display has already been given (Fig. 5). The records usually show echoes several miles high and several miles wide. Most of the displays used have actually tended to exaggerate the height; as Byers and Braham [38] point out, the maximum vertical and horizontal dimensions of most showers are about equal.

Early quantitative studies of shower development were made by reducing the receiver gain in a number of steps [39]. In this way contours of constant echo intensity can be drawn. For Fig. 11 this method has been used by the present writers to record the development of an echo during the first few minutes after it became detectable. Atlas [40] has elaborated a method for automatic contouring which is finding increasing application as quantitative aspects of radar observations are becoming more important.

Systematic measurements of echo heights and temperatures have been made by the Thunderstorm Project in 1947 (analyzed in detail by Battan [41]), by Workman and Reynolds [42], and by R. F. Jones [43]. Individual case studies, supported by balloon or in-flight observations, have been made by many other observers including Hilst and MacDowell

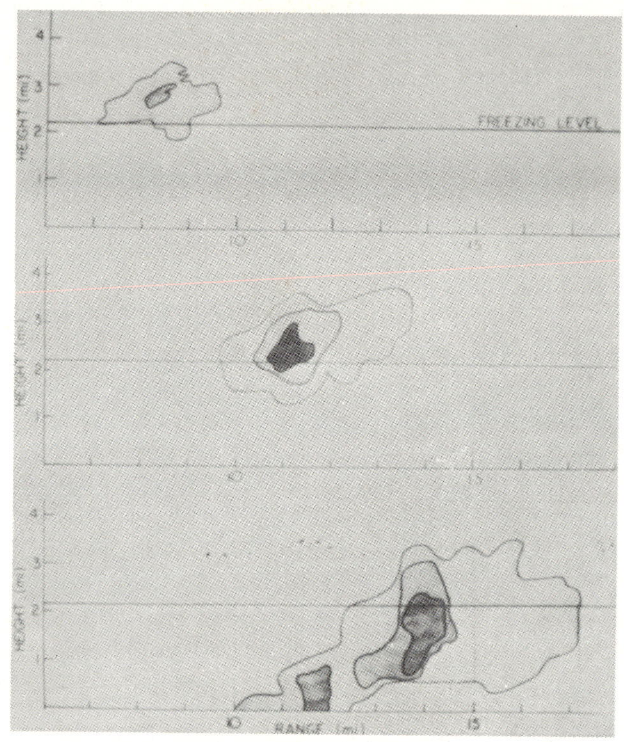

FIG. 11. Contours of M, the liquid water content, for a developing shower echo, derived from echo contours as the gain was reduced in successive steps. Contours shown are 0.1, 0.3, and 1 gm/m³. (Photograph, Stormy Weather Group, McGill University.)

[44], Bowen [21] and E. J. Smith [45]. Results are most significant where it is possible to follow a convective cell through its entire life cycle. Workman and Reynolds [42] made photographic and radar observations of individual cells from the stage when the cumulus tower had barely projected above the level of surrounding cloud. They attempted a composite description, derived from twelve such life cycles in New Mexico. The echoes first appeared at a height where the temperature was $-10°$ C (about 22,000 ft); the echo tops rose to around $-29°$ C in about 12 min and then subsided in another 12 min, falling more or less to the height they started from. This picture of a cell, typical of the arid

Southwest of the United States, does not seem to be applicable in the more northern regions reported on by Byers and Braham [38] and by Battan [41]. In Ohio, shower echoes were first observed with their tops at temperatures ranging from +14 to −16° C, with about 60% of the initial echoes occurring at temperatures warmer than 0° C. The 123 echoes so observed grew rapidly upwards at rates up to 800 ft/min, while the echo base was descending at rates up to 1500 ft/min (Fig. 12). These rates of motion seemed to depend on the initial height of the echo. In general, lower and warmer initial echoes grew more rapidly, and to a

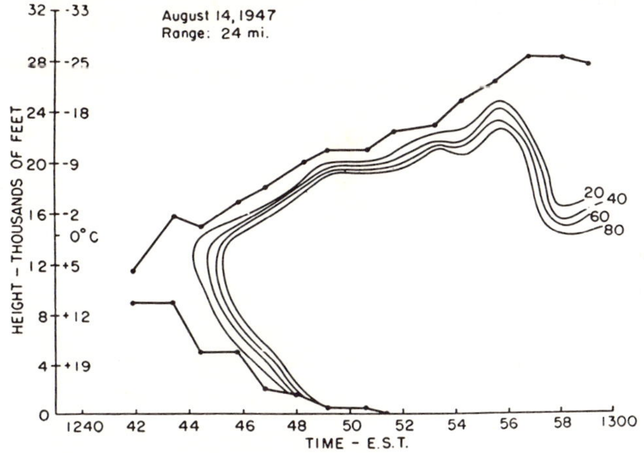

Fig. 12. Example of vertical growth of a storm echo in time showing contours of constant brightness on the radar scope (arbitrary units). (From Battan [41].)

greater extent. Although rates of rise of the tops may be explained by the updraft, the descent of the bases is probably faster than can be easily accounted for by the known terminal velocities of the particles and the observed downdrafts. Battan inclines to the view that the descent of the echo is due, at least in part, to drops growing to detectable size in the successively lower layers of the cloud. A study by Rigby, Marshall, and Hitschfeld [46] shows that the rates of vertical motion of the echo edges depend critically on the sensitivity of the radar. High rates of fall of the echo base may be anticipated if the core of the storm has an intensity appreciably above the minimum detectable.

The high temperatures associated with many of the echoes analyzed by Battan have caused him to assume a mechanism of precipitation not involving the ice phase. It is still uncertain, however, exactly how drop growth in the early stages can be sufficiently rapid by coalescence alone. Similar conclusions were drawn by Jones [43], Smith [45] and Bowen [21] on the basis of their studies. In addition, Bowen [47] has attempted a simple theory of droplet growth on the assumption of unit coalescence efficiency, that is, two droplets whose trajectories intersect were assumed always to coalesce.

Although Battan's radar data did not allow him to draw accurate intensity contours, a densitometer analysis of the photographic radar records has, in eight cases, revealed the existence of a sudden drop of the most intense portion of the echo when the cell has reached its maximum development (Fig. 12). Battan suggests that "the liquid drops grew and were carried upwards to the level where nucleation took place. When the cloud top reached this level, ice crystals formed in the upper regions of the cloud and spread rapidly downward through the rest of the supercooled cloud causing the liquid drops to crystallize. High reflectivity may be assumed to be associated with predominantly liquid drops, the low reflectivity with the mainly frozen particles." If such a phenomenon could be verified, it would of course lend substantial weight to the mechanism of precipitation proposed by Battan. On the other hand, in view of the fact that the radar used was a 3-cm set, it might be asked whether the observed decrease in the density of the photographs might not be due to the increase in attenuation since the beam had to pass through more and more cloud as the echo rose.

Reynolds and Braham [48] have attempted to explain the different findings of the New Mexico observations of Workman and Reynolds and the Ohio observations of the Thunderstorm Project. They assume a deficiency of condensation nuclei in the Southwest; thus the supersaturated air would have to be chilled to the low temperatures required by the sublimation nuclei before the formation of clouds. It is thus suggested that whereas the precipitation in Ohio involved only liquid particles which had originated on the abundant nuclei present in industrialized areas, "the New Mexico observations do not exclude" a mechanism of particle growth initiated by ice. This is in agreement with a study by Braham, Reynolds, and Harrell [49] who found that on clear days the majority of clouds in New Mexico developed no precipitation (indicated by the absence of radar echoes) because of the scarcity of nuclei. On overcast days, on the other hand, the crystals developed by one cloud seemed to be able to "seed" others, leading to a greater incidence of precipitation echoes.

This idea of natural seeding has been further developed by Dennis [50] who showed from an analysis of RHI photographs that showers are often associated with snow trails (Fig. 13). He suggested that the snow might be effective either as a seeding agent in the supercooled portions of cumulus clouds, or by increasing the moisture content of the entrained air. The effect of the snow would be particularly noticeable when the cloud tops are at temperatures between $-5°$ C and $-12°$ C, where precipitation is unlikely to initiate without this aid.

On the PPI display, showers appear as small blobs of fairly uniform brightness having well-defined edges. Often these echoes appear in parallel bands or lines, each line being made up of many cells (see, for instance,

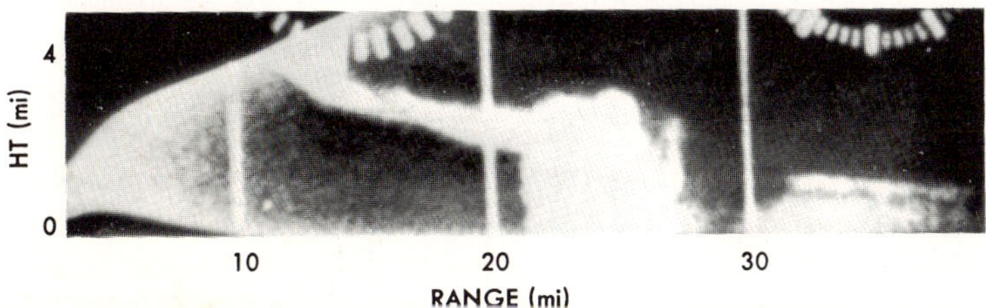

FIG. 13. An RHI display showing a shower echo extending up into a region of snow trails. (From Dennis [50].)

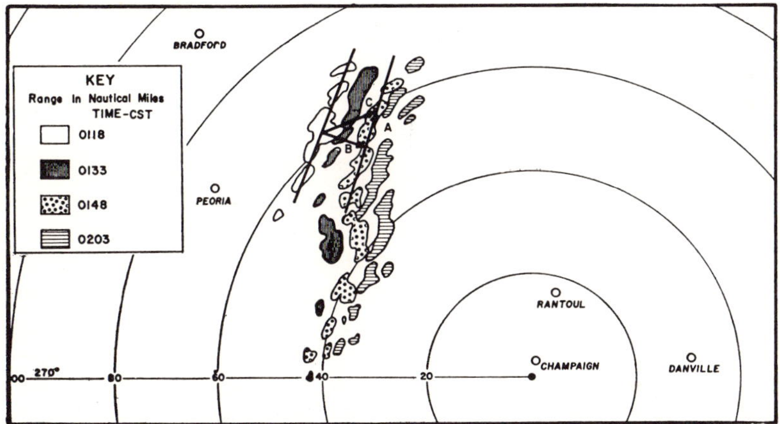

FIG. 14. Section of a PPI showing the position of a line of shower echoes at 15-min intervals. The displacement of an individual shower is given by the vector C, which is the resultant of the perpendicular displacement of the line (vector B) and displacement of the echo along the line (vector A). (Courtesy Illinois State Water Survey.)

Wexler [51, 52] and Austin [53] and Fig. 3). Such "squall lines" may be associated with a surface cold front, but occasional lines of showers occur which have no connection with a front. The lines may be up to 200 miles long. The motion of any one shower echo in these cases may be looked on as the resultant of two motions, one of the shower along the line, the other of the line perpendicular to itself. This is pointed out in Fig. 14 which is taken from Stout [54]. When showers are isolated, the direction of motion is reported without ambiguity. When showers form in an almost continuous line, errors may arise if the observer ceases to report individual shower motion and commences to report the motion of the line perpendicular to itself, ignoring the component of the motion along the line.

Swingle and Richards [55] reported a study of 44 storms in which lines of precipitation echoes occurred. They found the speed of motion of the lines to be quite erratic, and suggested that, in order to give any reasonably accurate prediction of the time of arrival of a given squall line, the speed of approach must be averaged over several half-hour intervals. In another paper, Swingle and Rosenberg [56] stressed the difficulty of forecasting thunderstorms and squall line activity from the conventional synoptic procedures. They suggested that accurate and detailed short-range (3 to 6 hours) forecasts are possible with radar information added to the synoptic data. They usefully pointed out that radar observations are of a scale that can be properly classified as neither synoptic nor micrometeorological, and so used the term "meso-meteorological."

Ligda [57] found greater consistency with synoptic data in studying the motion of *individual* shower echoes in some 800 hours of PPI records. He found that the observed velocity of the echoes correlated best with the geostrophic wind at the 700-millibar level (Table I) which in New

TABLE I. Correlation coefficients between precipitation area and geostrophic wind velocity (from Ligda [57]).

	850 mb	700 mb	500 mb
Speed	0.49	0.81	0.58
Direction	0.73	0.96	0.87

England is usually not far from 10,000 ft. Usually he observed the same direction of motion for all the echoes on a PPI screen of 150 mile radius. Occasionally cases of departure from parallel motion were found, but even in these cases the direction of motion was in general agreement with the notion of a 700-millibar steering level (Fig. 15).

A summary of the detailed weather information that can be obtained by a meteorologist from the study of a PPI display was given by Stout [54] at the 1953 Radio Meteorology Conference. He pointed out that the radar can show the experienced observer changing synoptic features, such as the approach or passage of different types of fronts and squall lines. A change in the direction or speed of any echo may reveal a change in steering winds. Varying echo movement in space may be used to determine the location of ridge and trough regions. It is possible that information about turbulence in cumulus clouds may be determined from the PPI echoes, for it is found that projections from strong, large echoes (Fig. 18) as well as steep contour gradients (when gain is reduced successively) are associated with destructive winds. Quite apart from the precipitation echoes themselves, one of the "nuisance features" of radar can be put to effective use. Conditions of low-level stability produce "anomalous propagation": the radar beam is abnormally bent and reveals an array of

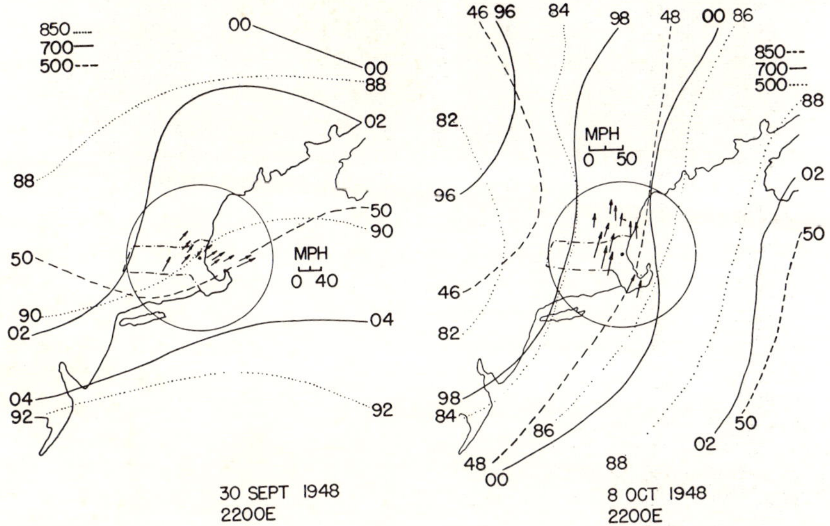

FIG. 15. Examples in which the direction of echo motion varies across a PPI display of 150 mile radius (circle). Vectors represent 1-hour displacements and show the agreement with the flow at the 700-millibar level (solid contours). (Contributed by Weather Radar Research, MIT.)

ground echoes not normally visible. Observation of such echoes then reveals the existence of low-level stability.

2.4. Hurricanes

The radar echo of a hurricane on a PPI display has characteristics which allow it to be easily identified (Figs. 16 and 17). The first such pictures to be published were those presented by Maynard [9]. Since that time, radar has been widely used in the tracking of hurricanes. One active group is the Florida Engineering and Industrial Experiment Station [58]. They observe the spiral bands of rain echo to move slowly around the center of the hurricane. The cells in the bands move along the bands and into the center (in a clockwise direction in the Northern Hemisphere). At least one commercial enterprise (Dow Chemical Co., at Freeport, Texas) [59] has installed a radar for hurricane warning and tracking in the hurricane zone on the Gulf of Mexico. Airborne radar has also been used to provide early warning of hurricanes before they

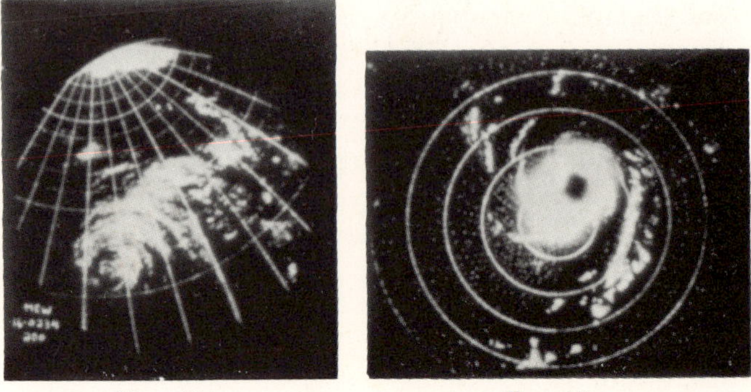

Fig. 16. Fig. 17.

Fig. 16. Off-center PPI display showing a hurricane. Range markers 10 miles, azimuth markers 10 degrees apart. (Obtained by D. Atlas, USAAF at Orla Vista Florida, September 1945.)

Fig. 17. Standard PPI display, with 20-mile range markers, showing a hurricane with the center of its "eye" about 20 miles northeast of the radar. (Official photograph, U.S. Navy.)

are within range of land based radars, and it has proved a generally useful tool in hurricane reconnaissance flights.

2.5. Tornadoes

Up to 1953, there had been some doubt as to whether a tornado would be detected by radar because of its small horizontal extent and because of masking by the thunderstorm activity in which the tornadoes develop. In April 1953, however, radar pictures of a tornado were published by Stout and Huff [60]. These photographs, a few of which are reproduced in Fig. 18, did reveal an echo with a distinctive shape, and with a motion

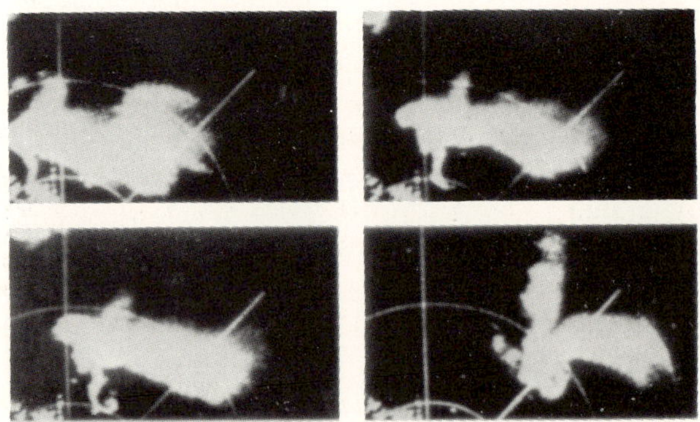

Fig. 18. Sections of PPI displays illustrating development of a tornado over Illinois on 9 April 1953. Top pictures are at 1705 and 1713 hours; below at 1716 and 1738 hours. Range markers are 10 miles apart. (Courtesy Illinois State Water Survey.)

differing from that of other nearby echoes. Great interest was aroused by these pictures, and by the radar observation of other tornadoes during 1953 (e.g., records obtained by J. C. Freeman at the Texas Agricultural and Mining Institute, and R. Blackmer at the Massachusetts Institute of Technology). In January 1954, Freeman's observation of a tornado by radar [61] was the first indication of the tornado's existence. He obtained a set of 50 time-spaced pictures of the radar display. There is a striking similarity between this sequence and that obtained by Stout and Huff in Illinois. The comparison indicates that not only the pattern of the appendage but the development in form with time of both the tornado and the parent storm possess characteristics by means of which a tornado may be recognized.

The tornado observations at Massachusetts Institute of Technology in 1953 revealed characteristic tornado pattern only when the sensitivity of the high-power equipment was reduced. The reliable location of tornadoes by radar will apparently require the development of careful techniques. Special accessory equipment to reveal the high velocities or high relative velocities may prove helpful. The detection of incipient tornado-spawning thunderstorms is perhaps of more practical importance than the location of the tornadoes themselves. Toward this end, observations in vertical section will undoubtedly be useful. Jones [62] has followed the development of such storms by "sferics," the radio-location of lightning. It is in the study of tornadoes and tornado-spawning storms that the combined operation of radar and sferics is most likely to prove its worth.

2.6. Lightning

Microwave radar echoes produced by lightning discharges were first observed in the Panama area during the last year of World War II. By visual observation, these transient echoes were correlated with lightning and, as pointed out by Ligda [63], they were true echoes since they occurred at the same range on a number of range sweeps of the radar display. At the moment it is not known whether the echo from lightning is returned by a sharp change of dielectric constant caused by a large and abrupt difference of temperature, or whether the echo results from scattering of the radar beam by the ionizing processes of the lightning discharge. Browne [64] considered these two possibilities and rejected the former.

Existing observations suggest that lightning echoes are frequent at wavelengths of 50 cm and 10 cm, and much less common at 3 cm. It may be that the precipitation echo, being relatively more intense at 3-cm wavelength, usually obscures any lightning echoes.

Marshall [65], using a 10-cm radar, found lightning echoes always within about one mile of the top of a shower echo, in regions of no, or at any rate undetectable, precipitation (Fig. 19). Miles [66], in Southern Rhodesia, also with a 10-cm radar, found echoes sometimes inside, sometimes outside the precipitation echo (Fig. 20). From the observations of

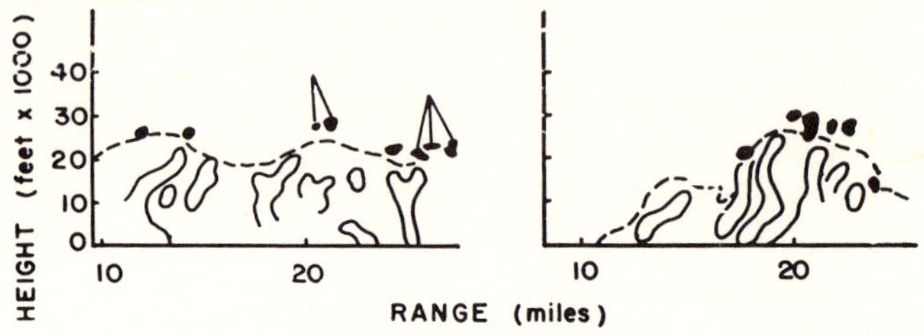

FIG. 19. Lightning echoes from two 5-min sequences are superimposed on a general sketch of the precipitation pattern during that sequence. Pairs and groups of three echoes that occurred simultaneously are indicated by joined straight lines. Broken lines indicate tops of echoes, solid lines, the densest portions. (From [65].)

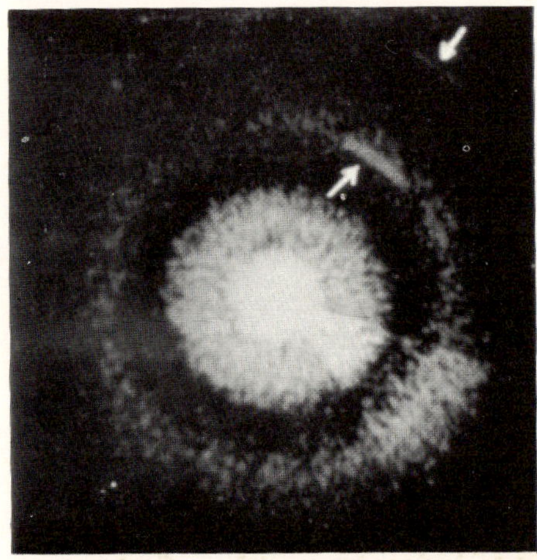

FIG. 20. A pair of lightning echoes simultaneously observed with a vertically pointing beam, so displayed that height is represented by radial distance, time by azimuth angle (6 seconds = 1 revolution). Widely distributed grainy echo is from precipitation. (From Miles [66].)

both these workers it seems generally true that the echoes come from regions at temperatures colder than 0° C. The occurrence of the echoes at freezing temperatures becomes significant in connection with theories of the separation of electric charge in thunderclouds.

Regarding duration, Ligda, Marshall, and Miles found the time to be most often not less than one-third of a second. However, Miles has reported durations of the order of one-thirtieth of a second for lightning echoes from distant storms. Hewitt [67] on the other hand found echoes even from relatively close storms to have durations as short as one-hundredth of a second. It seems that clarification of echo duration will have to await further observations. It is worth noting, however, that, since radar sensitivity falls off with range, distant echoes may appear to have shorter duration. This is apparent from a study of a pulse-integrator record presented by Ligda [63], on which lightning echoes appear to increase rapidly in strength for a small fraction of a second and then decay exponentially to the level of the background in from one to five seconds. Shorter duration echoes from the more distant lightning could thus be explained by the reduced radar sensitivity at greater ranges. In any case, it would seem worthwhile to study lightning echo duration as a function of range.

Although lightning echoes give promise of providing useful new information about the physical process of lightning, they do not provide a reliable lightning locator, not, at any rate, if a wavelength of 10 cm or less is used, as it is for best results with precipitation. "Sferics," the location of lightning by direction-finding on radiofrequency noise associated with the stroke, has, on the other hand, proved a useful technique for many years. A single sferics station reports only direction; two stations separated by a considerable base line are required for location. The usual procedure is to use a base line of a hundred miles or more, and to locate storms a thousand miles or more away by their electrical activity. Scaled down by a factor of ten, the sferics system should provide an excellent complement to radar observations of precipitation. It would be helpful then to avoid signals from greater distances, possibly by ignoring relatively weak signals, but more likely by tuning the receivers to a frequency not reflected by the ionosphere.

The promise of this combined operation has been agreed upon at radar weather conferences, but steps toward its institution have to date been quite modest. Swingle [68] has operated a single sferics equipment at the same location as a weather radar, and connected it to report by a radial line on the PPI display of the radar. Operation of a single receiver is justifiable, since, given the direction of the lightning, its range can usually be arrived at from the precipitation pattern without undue ambiguity. Correlation between the radar picture of precipitation and the radio picture of the lightning can also be achieved in vertical section if the storm is nearby. Workman and his colleagues [42] have reported the combination of visual, radar, and radio observations in this way.

2.7. Clouds

Due to the small sizes of the droplets in clouds, the reflectivity is about a factor of 10^6 less than in rain. Since the reflectivity is inversely proportional to λ^4, some of this loss can be recovered by going to shorter wavelengths. Less power is then available, but radars of wavelength 3 cm and shorter can detect clouds.

Most of the radars intended for cloud detection have used a fixed, vertically pointing beam in order to minimize the range. Wavelengths both somewhat greater and somewhat less than 1 cm have been used for cloud observations with some success. One difficulty with these short wavelengths is their extreme sensitivity to snow and rain, so that the cloud echo may be masked by even the weakest precipitation. Another is the great attenuation by rain and cloud and even, at 1.25 cm, by water vapor.

The Evans Signal Laboratory has made notable progress in radar cloud studies, having been engaged in this work since 1945 [69, 70, 71, 72]. Their early work was with a 1.25-cm set, which was abandoned for one at 0.86 cm when the latter showed much improved performance in cloud detection. Swingle [72] reported on a comparison between clouds detected by the 0.86-cm radar, and by hourly visual observations for 953 hours of daytime cloudiness from November 1951 to November 1952. The radar performance was inferior to that of the observer only in the case of fracto-stratus and fractocumulus type clouds (Table II).

TABLE II. Percentage frequency of cloud observations by radar and by weather observer, grouped by general height of the cloud base (from Swingle [72]).

Cloud group	Detection by: Radar only	Radar and observer	Observer only
Low clouds: (Cu, Cunb, Stcu, St, Frst, Frcu, Nbst)	11.5	71.2	17.3
Middle Clouds: (Ast, Acu)	46.6	39.4	14.0
High clouds: (Cist, Ci, Cicu)	51.5	38.2	10.3
All clouds	31.9	53.6	14.5

The Air Force Cambridge Research Center group under Atlas has also been active in radar-cloud studies. From simultaneous radar and cloud camera records they have made a comprehensive survey of characteristic echoes associated with the various cloud types [73]. Having identified cloud type visually, they derived the radar echo characteristics of that type so that it would be possible not only to detect but to classify the cloud types by radar observation. An attempt was made to determine just how many of the echoes, detected by their set, could be classified as true clouds. The records were examined for the presence of any virga-like trails or extensions below what appeared to be the steady echo base. It was found that 47% of the cases displayed extensions smaller than 2500 ft, and it is possible that many of these extensions may not have been virga but only irregularities in the cloud base.

As mentioned previously, considerable information about precipitation patterns has been obtained on the cloud set (Fig. 10) and given quantitative treatment.

2.8. "Angels"

For some years sporadic signals resembling point-target echoes of short duration have been observed, especially on the short-wavelength radars used for cloud studies. In 1952, Gould [74] reported frequent echoes occurring at relatively low levels, rarely above 8,000 ft (Fig. 21). Often

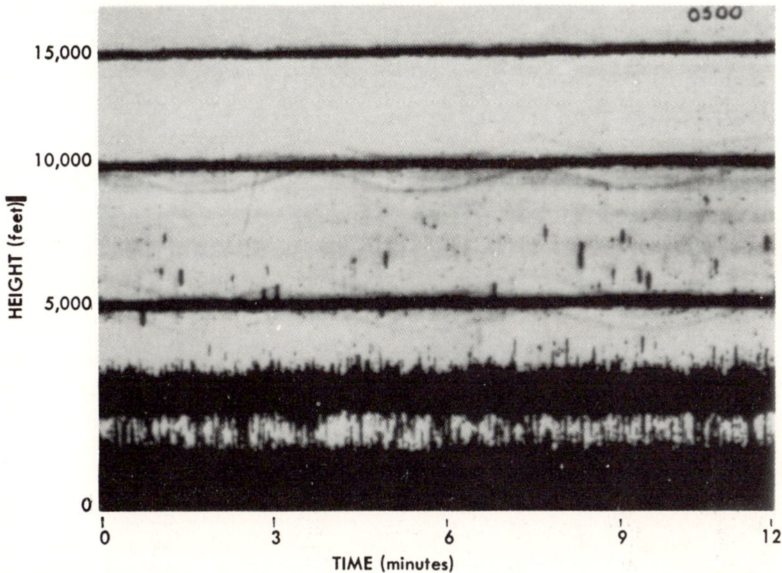

FIG. 21. A height/time display showing a horizontal layer of "angel" echoes extending from 1700 to 2500 ft. The lower band of echo from ground to about 800 ft is due to the transmitted signal. (From Gould [74].)

they were observed when the zenith was clear blue sky. It was suggested by Gerhardt in discussion of this paper that these echoes might come from sharp discontinuities in refractive index in the atmosphere, particularly since airborne refractometer measurements [75] had indicated much greater fine-scale variations in refractive index (of the order of 20–30 parts per million in a few hundred feet) than were previously supposed to exist.

More recently, Browne [76] has reported a similar type of echo, received from the zenith of a visually clear sky, with a 3.2-cm radar. The echo was a horizontal layer type of depth about 300 ft, and lasted about 30 min. A nearby radiosonde revealed that it was close in height to an anticyclonic inversion. Browne deduced that the echo must have been due to an exceptionally abrupt discontinuity in either temperature or humidity, or both. Although the sounding gave no evidence of so sharp a discontinuity, he recalled Friend's [77] argument from radar observations at still longer wavelength that transitions in the atmosphere may be far steeper than balloon sounding equipment has been capable of indicating.

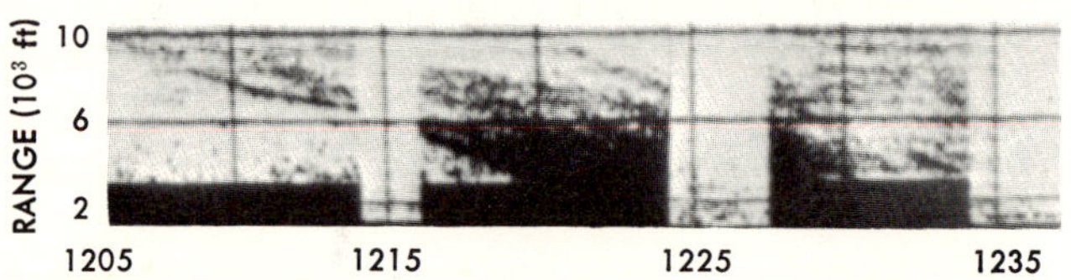

FIG. 22. A range/height display produced by a 1.25-cm radar directed horizontally out to sea from a point one-half mile inland. Atlas [79] interpreted this echo as being due to the seabreeze front. For three short periods (at 1215, 1225, and 1235), the beam was directed upwards; at 1235, the inversion reached the radar and a faint layer of echo can be seen overhead at about 1000 ft. (Courtesy, Geophysics Research Directorate, Air Force Cambridge Research Center.)

In 1953, Plank [78] reported a study of "angel" signals on 57 days found among 15 months' records obtained with a 1.25-cm zenith radar. Plank found that high temperatures, low wind velocities, high moisture content, and the absence of thick upper clouds represented the most favorable conditions for the appearance of "angels"; the echoes appeared in layer form in the vicinity of sharp moisture gradients aloft. He investigated the possibility that the echoes might be attributed to swarms of insects. After a careful comparison between echo behavior and atmospheric soundings and between echo behavior and insect behavior, he concluded that, although some of the echoes might be due to insect swarms, some definitely could not. On the other hand, practically all the echoes could be explained on the basis of refractive index discontinuities.

At the same time Atlas *et al.* [79] reported the appearance of angels on a seabreeze front (Fig. 22). The echoes moved in from the sea, and coincident with their arrival at the shore the temperature there dropped 2° C, the wind changed by 180 degrees, the relative humidity rose, and the refractive index (as measured on a microwave refractometer) rose rapidly. Although these data have not yet been fully analyzed, it appears that the angels arose in a region where the refractive index changed rapidly with distance. More precisely, they would be attributed to random inhomogeneities in this region of rapidly changing refractive index. One difficulty with the analysis is that, as the seabreeze front passed by, the refractometer gave a change of 30 N-units* in 45 min, whereas from the corresponding temperature drop and vapor pressure increase, the change was computed to be only 8.8 N-units.

It may reasonably be hoped that opposed to their modest nuisance value, angels may yet prove useful target elements in permitting the radar exploration of clear air, as first envisaged by Friend [77].

2.9. Summary

In summary, radar has revealed the pattern of precipitation in horizontal section over circular areas of radius 200 miles, revealing the life cycles of individual storms, their grouping both in intensive squall lines and in more extensive pattern, their individual motions, and the pattern of motion over the whole area. Synoptically, the area covered by a single station is rather small, leading to the coining of the expression "mesometeorological" to describe the scale of the observations. The introduction of radar networks, notably by the Air Weather Service of the USAF with their CPS-9 program, should extend the observations to synoptic scale. Meanwhile, single-station observations find practical application in adding specific detail, relatively local but up to the minute, to the more general information provided by normal synoptic methods, for the benefit of pilot briefing and short-range metropolitan forecasting. The range of a single station has proved sufficient for hurricane warning and tracking, and for the use of radar to complement in-flight observations in providing a study of the whole precipitation pattern of a cyclone.

Radar observation of precipitation has thrown much light on the precipitation mechanisms. Combining observations in horizontal and vertical section, it has been possible to derive the geometry of the rain shower and the thunderstorm in three spatial dimensions and time. On the time scale, observations seem to have been obtained within a minute of time zero, the time at which precipitation is initiated in a shower; findings regarding the point in the cloud at which precipitation is initiated and regarding the rate of initial development are of special interest.

Radar has provided the best evidence for shower formation that does not involve an ice phase either as a primary or a subsidiary part of the mechanism. It has also shown that continuous rain is essentially melted snow. It has shown that the mechanism of continuous snow formation usually possesses a marked macroscopic pattern, involving convection more often than it involves instability. The process appears to be con-

* $N = (n - 1)10^6$, where n is the refractive index.

tinuous in time rather than in space. It is providing useful evidence concerning the height and temperature of intensive snow growth.

At shorter wavelengths, the use of radar to delineate cloud formations has been developed. The cloud pattern as such tends to be masked by even light precipitation, but the vertical pattern of very light precipitation revealed by cloud radars can be used in practice to decipher the cloud pattern aloft, and is most useful in providing data for research considerations. Radar returns from clear air, reported by these cloud radars, have served more to excite scientific curiosity than to meet a practical need. Since those returns do contain information about inhomogeneities in cloud-free air, not now clearly understood, it seems reasonable to hope that their further study will be rewarding.

One final word about the study of pattern: before proceeding to discuss what we term "quantitative" radar observations, let it be noted that the precision of location and relative location with which the precipitation pattern is revealed on a radar screen is the most "quantitative" aspect of radar weather observations.

3. Theory and Quantitative Observations

3.1. The Radar Equation

The intensity of a radar signal from precipitation is related to the properties of the scattering particles and to the characteristics of the radar according to the equation

$$\overline{P_r} = \frac{P_t A_e h}{8\pi r^2} F k \Sigma\sigma \tag{1}$$

In this equation, which has been reported in several forms by many writers, differing only in detail (e.g., Kerr [8], section 7.1), $\overline{P_r}$ is the average value of the power received at the radar, P_t is the power transmitted by the radar, A_e is the effective receiving area appropriate to an extended target (between 0.5 and 0.7 of the geometrical area), h is the pulse length, and r is the range of the target. The quantity $\Sigma\sigma$ is the sum of the back scatter cross-sections of the scattering particles in unit volume. The factor F has not yet been explained;* it is inserted into equation (1) on empirical grounds. Careful measurements by Austin and Williams [80] have shown that F has a value about 0.2 for their SCR-15 10-cm radar. The factor k is the reduction in $\overline{P_r}$ due to attenuation by atmospheric gases, clouds, and rain. The exact evaluation of k is usually quite complicated. For this reason, quantitative measurements have generally

* See footnote in section 3-5.

not been attempted at attenuating wavelengths except near the front edges of the storms, where the assumption that attenuation was negligible (i.e., $k = 1$) was considered safe.

Equation (1) is valid provided that the cross-section of the radar beam is filled with target precipitation. If the target is smaller than the beam at range r, but would fill the beam at a shorter range r_0, the term $1/r^2$ in the equation needs to be replaced by r_0^2/r^4. When the beam is filled in one dimension but not in the other, this factor should be r_0/r^3.

The fundamental problem in applying equation (1) in weather observations is the correlation of $\Sigma\sigma$ with the characteristics of the precipitation. Although a complete discussion of this problem is beyond the scope of this review, we shall summarize the manner in which $\Sigma\sigma$ varies with the rate of rainfall, and shall indicate briefly how particle shape and dielectric constant may be taken into account. We shall also discuss the effect of attenuation and the fluctuating nature of the signal and give a brief description of attempts at interpreting the power of the echo in terms of the intensity of precipitation and the density of cloud.

3.2. Back Scattering from Spherical Particles

The simplest case is that of a spherical water drop or solid hail-stone; yet even it is far from simple. The exact theory of scattering from spheres has been worked out by Mie [4] in terms of two infinite series in α, where $\alpha = \pi D/\lambda$, with D the particle diameter and λ the wavelength. The evaluation of the coefficients in these series requires a good deal of computational effort; a few special cases have been worked out by Lowan [81].

A considerable simplification results when α is small, say less than about 0.13. Then, only the first term of one of the series needs to be considered in a form which is known as the Rayleigh approximation (see, for instance, Gunn and East [107a]). Here

$$\sigma = \frac{\lambda^2}{2\pi} 2\alpha^6|\kappa|^2 = \pi^5|\kappa|^2 \frac{D^6}{\lambda^4} \tag{2}$$

where $\kappa = (m^2 - 1)/(m^2 + 2)$, with m the complex index of refraction. This result expresses the important fact that σ is inversely proportional to the fourth power of the wavelength, and directly proportional to the sixth power of the particle diameter; and so involves the particle size in only one term. (These facts were brought out by Rayleigh in his discussion of optical scattering by the clear sky [3].) On account of the simplicity of equation (2), one useful way of expressing σ for all drop sizes is to use Rayleigh's form (2) and to allow—where necessary—for a correction term $\epsilon = \sigma_{\text{Mie}}/\sigma_{\text{Rayleigh}}$. In Fig. 23, for which Lowan's data were used, ϵ for water is plotted as a function of the drop diameter for three wavelengths. The curves show that the Rayleigh approximation without correction is rather good for values of D up to 1 mm when $\lambda = 0.9$ cm, up to 2 mm when $\lambda = 3$ cm, and nearly up to 5 mm when $\lambda = 10$ cm. The great difference in the appearance of the curves is partly due to the variation of the refractive index m with wavelength. For snow and ice, Rayleigh's approximation is valid for much greater particle diameters, but Rayleigh scattering is much weaker here than for water spheres of the same mass (the difference is 6.5 db at 3 cm).

There is one important extension that is amenable to exact treatment: the scattering by spheres surrounded by a concentric film of different

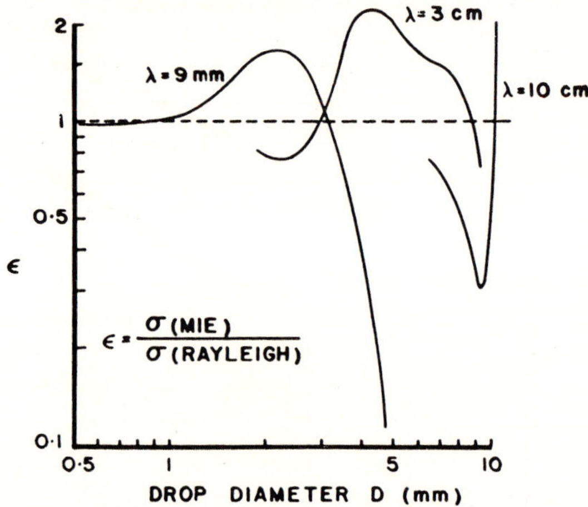

FIG. 23. Ratio of exact (Mie) to approximate (Rayleigh) back scatter cross-section of water spheres of diameter D at three wavelengths. (Calculated from Lowan [81].)

dielectric constant. The complete theory based on Mie's work for homogeneous spheres has been given by Aden and Kerker [82]; some calculations for the case of an ice sphere surrounded by a film of water have been reported by Kerker, Langleben, and Gunn [29], and more completely by Langleben and Gunn [83]. The results are complicated inasmuch as they cannot be represented in terms of a simple set of universal parameters; it thus seems impossible to extrapolate from the solution for one set of values of particle size and wavelength to any other set. But in general it appears that even a thin coating of water greatly enhances the scattering of the particle, increasing its effective dielectric constant practically to that of an all-water sphere. The same may be said about the attenuation, which is of particular interest, as ice particles (without water) do not attenuate at all. Labrum's [30] direct observations bear out these conclusions, at least as regards scattering.

3.3. *Assemblies of Particles*

Before the results referred to above can be incorporated into equation (1), it is necessary to sum the σ's for all the particles present in unit volume. This requires a knowledge of the distribution of the particles with size. Many observers have studied this problem for rain, particularly since the advent of radar; the definitive distributions are probably still those of Laws and Parsons [84], although some doubt exists as to

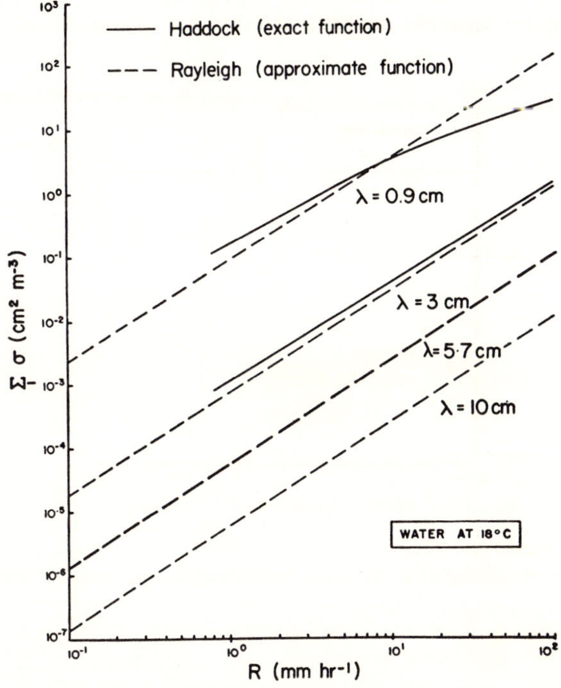

FIG. 24. Reflectivity of rain ($\Sigma\sigma$) as a function of R, the rate of rainfall, at four wavelengths. Solid lines: exact values calculated by Haddock [85]; dashed lines: approximate values calculated from Rayleigh's formula (2).

their general applicability in all cases. Using Laws and Parsons' results and Lowan's tables, Haddock [85] has worked out $\Sigma\sigma$ from Mie's formulas for σ. The results are shown as a function of the rate of rainfall in Fig. 24 (solid lines).

These accurate results may be compared with calculations performed on the basis of Rayleigh's formula (2), in which $\Sigma\sigma$ is proportional to ΣD^6, a quantity which, following reference [86], is often denoted by Z. The value of this parameter has been related to R, the rate of rainfall, by a number of writers, including Wexler [87], Marshall and Palmer [88], Hood [89], Best [90], Twomey [91], and Blanchard [92]. More studies may be expected in the future, as more detailed and more extensive drop-size measurements continue to be made, some of them by means of ingenious automatic methods (e.g., Bowen and Davidson [93], and Mason and Ramanadham [94] on the ground; and Adderley [95] aloft). In addition, several observers are photographing individual drops to study their shapes (Jones and Dean [96], Magono [97]).

The Z/R correlation is usually given in the form $Z = aR^b$, but different writers have assigned different values to a and b. When all the equations are plotted on the same scale, it is found that the loci of the majority of them are in remarkably close agreement. Only one of them will therefore be quoted

$$Z = 200R^{1.6} \text{ mm}^6 \text{ m}^{-3} \tag{3}$$

This relation, where R is in mm/hr, is the present writers' slight revision of the equation given in [88], and the error of estimate in R for a given value of Z is thought to be better than a factor of 2. A combination of equations (2) and (3) leads to the dashed lines of Fig. 24. In the case of 5.7- and 10-cm wavelengths, where calculations like Haddock's are not available, the accurate locus and the Rayleigh approximation may be assumed to agree closely. An important conclusion to be drawn from this figure is that whereas Rayleigh's approximate form for σ is not applicable at the largest drop sizes encountered, the Rayleigh approximation for $\Sigma\sigma$, the radar reflectivity of rain, is generally adequate at wavelengths of 3 cm and longer.

In the case of cloud the Rayleigh approximation for σ may always be used at centimeter and millimeter wavelengths without correction, because cloud droplets are always less than 100 microns in diameter. The reflectivity $\Sigma\sigma$ is then proportional to $Z = \Sigma D^6$. The significant meteorological cloud parameter is M, the liquid water content, which Atlas and Boucher [98] were able to relate to Z. On the basis of an analysis of some 100 cloud observations they concluded that cloud distributions may be described adequately by means of an equation with a single parameter, for which they suggested the median droplet diameter. Although further analysis indicated the desirability of distinguishing between at least three types of cloud, an over all regression equation

$$Z = 0.0292M^{1.82} \text{ mm}^6 \text{ m}^{-3} \tag{4}$$

was arrived at. Here the standard error of estimate in M, which must be measured in gm/m³, is about 50%. A comparison of this equation with equation (3), using average values for M and R shows that the reflectivity of cloud is of the order of 60 db less than that of rain.

Although the problem of spherical ice particles coated with a film of water has proved amenable to exact solution, the even more important problem of scattering from particles of shapes other than spherical has defied exact analysis. Two independent approximate treatments, more or less appropriate to the situation in radar meteorology, have been attempted (Labrum [32] and Atlas, Kerker, and Hitschfeld [99]) and a limited set of laboratory measurements has been reported by Labrum [30]. The two theoretical papers, which were in substantial agreement, dealt with ellipsoids of revolution; next to spheres these are the most tractable shapes. It was recognized, of course, that the actual shapes of the particles—flattened raindrops, hail stones, snowflakes, and particles in the process of melting—were much more complicated indeed. Whatever the shape, the assumption of axial symmetry was probably justified. Both theories were first-order approximations; the problem was reduced to one in electrostatics, which is justified only when the scattering particles are small relative to the wavelength. (This was precisely the earlier approach of Gans [100], who generalized Rayleigh's approximation to ellipsoids.) The results gave the reflectivity $\Sigma\sigma$ as a function of the ratio of axis to diameter for constant particle volume. Both randomly oriented particles and particles with preferred orientations were studied. In every case, $\Sigma\sigma$, the reflectivity, increased as the eccentricity of the principal elliptical cross-section was increased, with a similar statement applying to the attenuation and the depolarization of the echo. Curves of $\Sigma\sigma$ and the depolarization for randomly oriented water and ice spheroids are shown in Figs. 25 and 26 (taken from [99]). In addition, the case of ice spheroids coated with confocal water films has been treated [32]. Some results are shown in Fig. 27. In the case of a sphere, these results agree well with those for small spheres previously referred to [83]. They agree also, at least qualitatively, with such direct laboratory measurements of scattering as have been reported [30].

It is clear that the increase in scattering and attenuation due to non-spherical shapes has an important bearing on theories of the bright band (see section 2.2). Although at least an approximate knowledge of the scattering and attenuation by melting particles is now available, the essential obstacle to a satisfactory theory of the bright band remains our ignorance of the distributions of the particles with size and shape during melting. The authors just mentioned, as well as Austin and Bemis in 1950 [31], referred to the depolarization produced by shaped particles. Measurements of the component of the radar echo which is polarized perpendicularly to the transmitted beam would seem to offer a means for studying the deformation of the particles. Such determinations have been described by Browne and Robinson [101]. Recent progress with specially modified 3.2-cm equipment at the British Radar Research Establishment (formerly TRE) was reported in 1954 by Hunter [102],

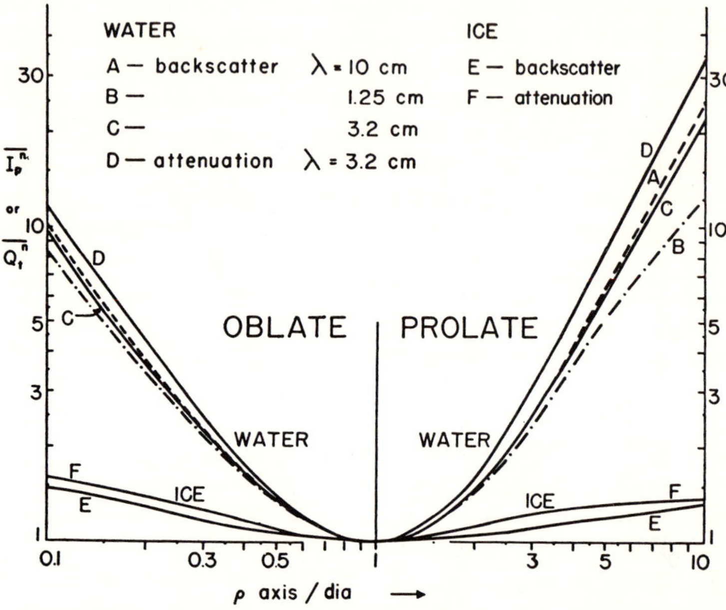

FIG. 25. Reflectivity and attenuation by randomly oriented water and ice spheroids, relative to values for equivolumic water and ice spheres. (From [99].)

who obtained good accuracy by measuring the ellipticity of the precipitation echo when the transmitted beam was circularly polarized. He reported the ratio of the orthogonal circular components, a quantity about twice the depolarization ratio for a plane polarized beam. Table

TABLE III. Depolarization ratios (in db) in precipitation echoes calculated from [102], with effective values of the ratios of axis to diameter for randomly oriented oblate (ρ_0) and prolate spheroids (ρ_p) which would give rise to the observed depolarization.

	Rate of precipitation								
	"Gentle"			"Normal"			"Heavy"		
Type of precipitation	db	ρ_0	ρ_p	db	ρ_0	ρ_p	db	ρ_0	ρ_p
Snow	29	0.75	1.35						
Melting band	23	0.76	1.22	20	0.66	1.32	18	0.59	1.5
Rain	38			30	0.85	1.10	26	0.82	1.17

III summarizes Hunter's data, to which we have added the effective values (ρ_0 and ρ_p) of the axis-diameter ratios, which on the basis of Fig. 26 would be required by the observations. These deformations are seen to be appreciable, except in the case of "gentle" rain; in the melting band

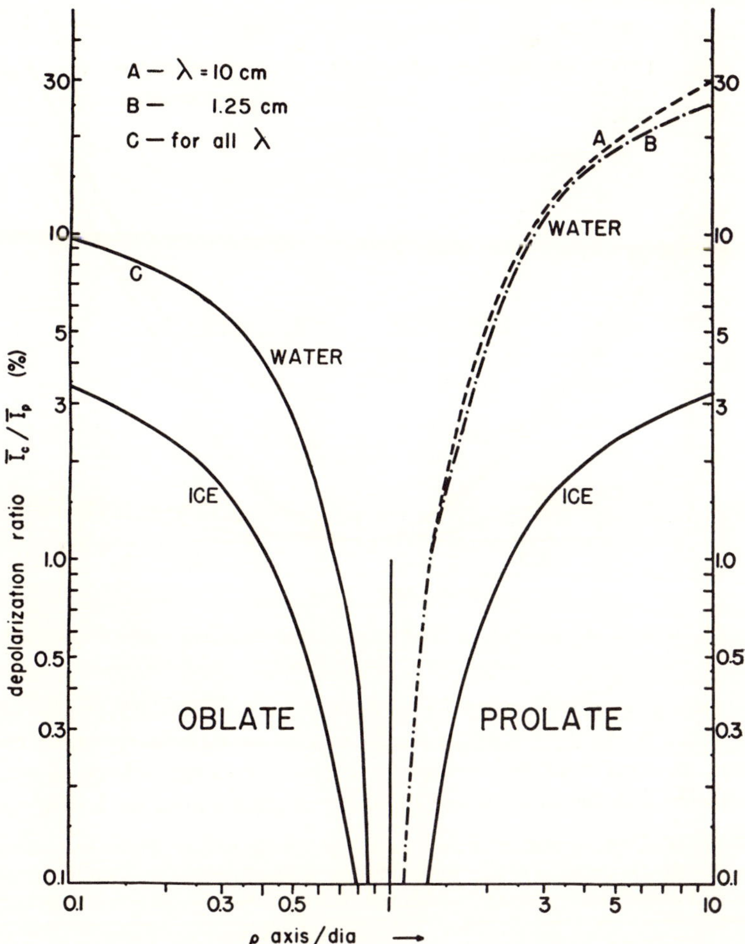

FIG. 26. Ratio of cross-polarized component of echo power to parallel-polarized component for randomly oriented water and ice spheroids. (From [99].)

they are always greater even than those in heavy rain. The values of the ρ's listed may be used to estimate the increase in the scattering and attenuation to be expected. An inspection of Fig. 25 shows that the increase in either of these is rather less than a factor of two in all cases, and is negligible for all precipitation outside the melting band.

Dry snow above the bright band consists of ice crystals, either single or aggregate. Although an aggregate flake consists of a mixture of air and ice of low mean density d, it has been shown [103] that $|\kappa|/d$ is constant as the particle density varies. (For the definition of the dielectric factor κ see equation (2).) This means that $\Sigma\sigma$ for snow can be computed

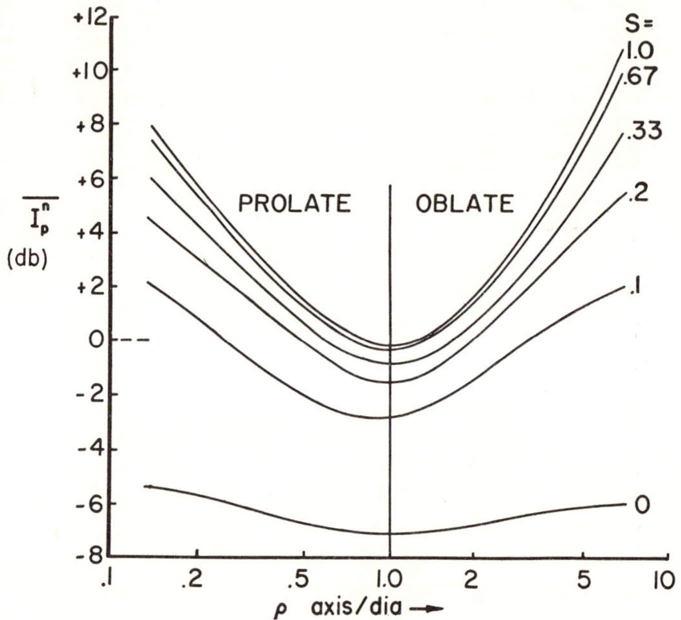

FIG. 27. Reflectivity (in decibels) of randomly oriented ice spheroids coated with a confocal film of water, relative to reflectivity of an equivolumic water sphere. Melting ratio, S = (volume of liquid water)/(total volume). (From [32].)

by catching a sample of snowflakes, melting them, measuring ΣD^6 for the resulting drops, allowing for the dielectric factor and, of course, for the terminal velocity. As shown in Fig. 25, the effect of the shape of (dry) ice particles on $\Sigma\sigma$ is generally unimportant, never as great as a factor of two. Marshall and Gunn [103] analyzed measurements by Langille and Thain [104] and obtained an empirical relation for snow

$$Z = 200R^{1.6} \text{ mm}^6 \text{ m}^{-3} \tag{5}$$

identical with equation (3) for rain, where R is measured in millimeters of water per hour. Since the dielectric factor of ice is about one-fifth that of water, radar echoes from snow have one-fifth the power of those from rain of the same rate of precipitation.

3.4. *Attenuation*

Equation (2) shows that the power back scattered to the radar is inversely proportional to the fourth power of the wavelength. This fact has been a major reason for radar weather observers turning to shorter wavelengths. The resulting gain in sensitivity over that at 10 cm which, other things being equal, amounts to 21 db at 3 cm and 40 db at 1 cm, is certainly desirable, particularly in work on snow and cloud. Another great advantage of the shorter wavelength is the increased resolution: for an antenna of given size the cross-section of the beam is directly proportional to the wavelength. Conversely, for a given resolution, a shorter wavelength means a proportionately smaller and lighter, and therefore more easily scanned, antenna.

But combined with these important advantages, a decrease in wavelength means an increase in attenuation. This quantity enters into equation (1) through the factor k, which in turn consists of several terms according to the equation

$$10 \log k = 2 \int_0^r (g + c + p)\, dr \tag{6}$$

where r is the range from the radar. The three terms in the integrand represent the attenuation (in db/mile, one way) by atmospheric gases, cloud, and rain, respectively. No terms are included for ice or snow, as their attenuation is negligible at all wavelengths. The gas term g, which depends on the atmospheric pressure and the density of the water vapor is usually not important. The cloud and rain terms may be written in the forms

$$\begin{aligned} c &= K'M \\ p &= KR^{\alpha} \end{aligned} \tag{7}$$

where M is the cloud density in gm/m^3 and R is the rate of rainfall in mm/hr. Values of K', K, and α, which depend only on the wavelength, are listed in Table IV. This table suggests that, at $\lambda = 10$ cm, the attenua-

TABLE IV. Attenuation constants calculated [106] from data given in [107a]. (Units of these quantities were chosen to fit equations (6) and (7), where r is measured in miles.)

	Wavelength, λ (cm)			
	0.9	3.2	5.7	10
K'	1.6	0.137	0.041	0.0144
K	0.25	0.0144	0.0047	0.00049
α	1.0	1.3	1.1	1.0

tion is quite negligible under all conditions; it is not quite negligible at 5.7 cm, and it is serious at 3.2 and 1 cm.

Attenuation is a nuisance where it occurs. It works against the enhanced *resolution* of the shorter wavelengths by seriously distorting the precipitation pattern appearing on the radar screen. This has been brought out by Atlas and Banks [105] who studied in detail a representative number of model storms. The same authors also showed [35], that under certain conditions, the shadows cast by echoes on attenuating wavelengths may give rise to spurious layers in echo regions.

Attenuation renders quantitative rain *measurements* difficult or impossible. Although Hitschfeld and Bordan [106] have devised a relatively simple procedure to correct for attenuation, provided the performance of the radar is known precisely, their general conclusion is that rain measurements should not be attempted on attenuating wavelengths. They did show, however, that when the radar set is calibrated against a rain gage, the correcting procedure leads to good precision in the measured values of rainfall at points where the attenuation due to intervening rain is smaller than at the gage. Attenuation also works seriously against the greatly enhanced *sensitivity* of the shorter wavelengths. Hitschfeld and Marshall [107] calculated the amount of intervening rain through which radars of different wavelengths can detect a given precipitation target. They concluded that, for observations *through* rain under operational conditions, the range to be expected on wavelengths 3 cm or shorter would probably be inadequate.

In concluding this section, a positive aspect of attenuation, recently elaborated by Atlas [98], should be mentioned. Attenuation depends on cloud density or the intensity of the intervening rain, or both. Hence a measured value of the attenuation would provide additional information about these meteorological conditions. No use has been made of this possibility to date, however, probably because of the difficulty of making measurements at the same time on two radars of different wavelengths.

3.5. Fluctuations

The amplitude (signal voltage) from precipitation at any instant is the vector sum of the amplitudes of the signals returned to the radar by all the particles in the contributing region at that instant. Since the particles are randomly spaced, the phases of their contributions are random, and the scattering is "incoherent." Under these conditions the *average* power of the echo is equal to the sum of the powers returned by the contributing particles (Rayleigh's well-known theorem of the "random walk" [108]). This average received power is the quantity $\overline{P_r}$ which has been related to $\Sigma\sigma$, the reflectivity of the precipitation, in equation

(1). The individual particles are, of course, continually in motion relative to each other and relative to the radar. To the extent that these motions are random, the instantaneous value of the echo power from a given target region fluctuates randomly about its mean. It follows that single-power readings are not significant, and that averaging of several power values is always necessary when weather observations are made.

The rate at which independent data become available depends on, among other things, the rate of scanning of the beam and on the relative

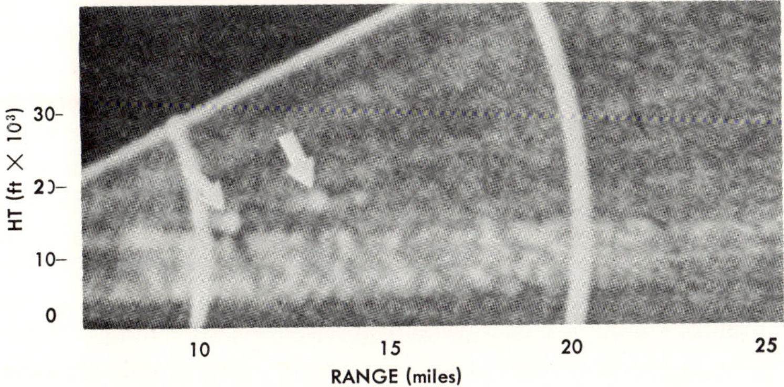

FIG. 28. Vertical section through cold frontal continuous precipitation. Note grainy texture of echo due to insufficient averaging. Time of exposure for picture was equal to time for one scan of beam. (Arrows point at lightning echoes.) (From [65].)

motion (henceforth called "shuffling") among the particles. For a stationary beam, the echo power returned from a given contributing region changes to an effectively new and independent value after a time T, which has been measured to be of the order of 20 msec (Fleisher [109, 110]) on 3-cm radar and of the order of 4 msec at 1.25 cm (Bartnoff, Paulsen, and Atlas [111]). Hence if the trace requires a time longer than this time-to-independence T in sweeping through its own width, and provided the pulse repetition period is shorter than T, the signal observed at each point on the display will be an *average* of several independent signals. In this way some averaging is achieved in ordinary radar operation thanks to the phosphorescence of the tube. Yet the results may still be grainy; Fig. 28, for instance, shows a texture which has no meteorological significance. Moreover, the detailed pattern of this texture varies haphazardly from one picture to the next. More, and better controlled, averaging may readily be obtained by reducing the rate of scanning (see remarks in section 2.2 about zenith-pointing radars), or by longer exposures. Figure 4 shows the fine grain resulting when the photographic record is a superposition of 10 complete scans, taking about 10 sec. Such a procedure is, of course, costly in time; in the above example, it takes ten times longer to obtain a smoothed vertical section than a coarse-grain one. Even more extravagant in time, although highly precise, is the "pulse integrator" developed at MIT [112, 113] in which any desired number of signal amplitudes from a *single* contributing region are averaged electronically, the average appearing as a meter reading. More ambitious schemes, such as pulse integration over the whole face of the display tube, have been proposed, but because of the complexity of the electronic gear involved they have not yet been realized.

The general mathematical theory of fluctuations has been known for some time (see, e.g., the review in Kerr [8]). Specific applications to radar weather have been made in some detail by Marshall and Hitschfeld [114] and by Wallace [115]; experimental checks for some of the basic ideas were recently secured by Bartnoff, Paulsen, and Atlas [111] and by Austin [116]. A finding of general importance is that the precision attainable in averaging independent power values increases with the square root of the number of data available. Averaging several independent values is, however, difficult and costly. It is much easier to *count* the number of signals falling above one or more suitably chosen thresholds, and such a procedure yields the desired mean only slightly less rapidly than averaging. When more independent data are needed (for high precision and good resolution) than are supplied by the natural shuffling of the precipitation particles, the number of independent data from a given region can be increased by transmitting consecutive pulses at slightly different radio frequencies, or by receiving the echoes at separate antennas.*

Because the nature of the fluctuations depends on the motion of the particles, observations of the fluctuations should give information about particle motion, notably turbulence. This was the basic idea behind "Rasaph," a device constructed at MIT [117, 118] for obtaining the (audio) frequency spectrum of the fluctuations. Useful results were obtained, but work with the instrument was discontinued when essential errors introduced by the circuits were recognized. These errors could

* Dr. W. E. Gordon of Cornell University, studying radio propagation by scattering from atmospheric inhomogeneities, has noticed that the energy received is not proportional to the effective antenna area when the antenna size is greater than the "correlation distance," i.e., the distance between two points at which the received signals have zero correlation. This distance depends on the size of the region in which scattering occurs. (This material has been submitted to *Proc. I. R. E.* for publication.) In the weather radar case the correlation distance is proportional to the diameter of the single antenna. Dr. Gordon has estimated in private discussion that the response in this case will be smaller than that to a plane wave by a factor of approximately 3. It seems likely that more careful consideration will establish this effect as the cause for the discrepancy between the measured intensity of the weather echo and that calculated on existing theory, i.e., the cause for the factor F in equation (1).

have been overcome only by unduly lengthening (from the original 1 min) the time allotted for the analysis of the echo from each contributing region (Fleisher [119]). Work on the radar signal spectrograph led, however, to a greatly improved understanding of the relationship between particle shuffling and echo fluctuations, and to the realization that more

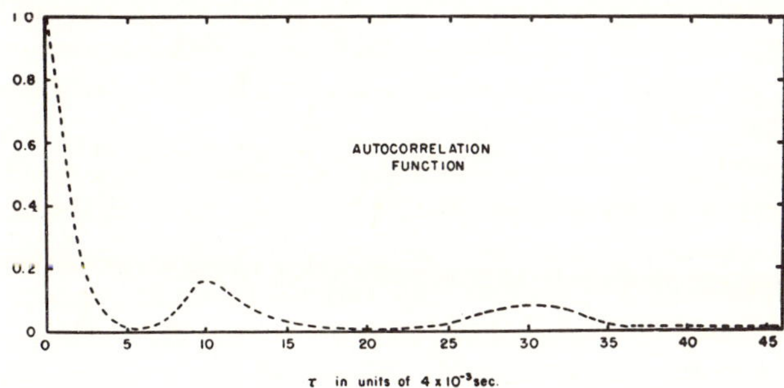

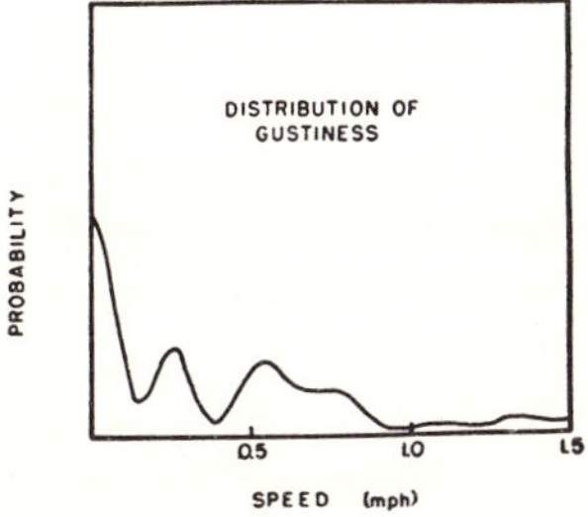

FIG. 29. Sample plot of autocorrelation function of 3-cm weather echo, and distribution of gustiness deduced from it. (From Fleisher [109, 110].)

direct turbulence information could be obtained from an autocorrelation study of the signal fluctuations.

Fleisher [109, 110] has reported a number of plots of the autocorrelation function $\rho(\tau)$ obtained from weather echoes of a CPS-9 radar. These curves, an example of which is reproduced in Fig. 29, show how the signal value at any time is, on the average, related to the signal at a time τ later; $\rho = 1$ means that the signals are identical, $\rho = 0$ means that they are independent. From these plots of ρ, Fleisher deduced the "distribution of gustiness" in the contributing region. This gustiness is, of course, a significant description of turbulence. Fleisher [109] pointed out, however, that more detailed turbulence information could be obtained from the weather echo, but excessive amounts of computational labor, and presumably valuable circuit time, would be required.

For practical purposes, say those of aviation, it is questionable whether more *detailed* measurements about turbulence are really needed. What may satisfy the practical requirements much better is a measure of turbulence rather than a complete distribution, say one number for each contributing region, provided this index can be gathered and displayed for a vast area in a short time. The ideal would be a turbulence map every time a PPI or RHI weather display is painted. Some work in this direction is now in progress. At the Massachusetts Institute of Technology efforts are being made to develop a rapid method of obtaining a single index of the rate of echo fluctuations. At McGill University an inquiry is being made to see whether a strong correlation can be expected between the fluctuations and the turbulence. It must be realized that, besides turbulence, differential settling of the particles, wind shear, receiver noise, and the scanning of the beam contribute to the fluctuations. It is too early to be sure that other factors do not mask the effect of weak or average turbulence. On the other hand, when the turbulence is great, the fluctuations may become so rapid as to be unmeasurable at any practicable pulse repetition frequency.

Other means of studying turbulence by radar have been suggested. From combined aircraft and radar observations, R. F. Jones [120] and Tolefson [129] found that the turbulence in cumulonimbus clouds was greatest near sharp echo discontinuities, where the gradient in echo power was a maximum. If these echoes are studied quantitatively by intensity contours (e.g. [39] or [40]), the spacing of the contours would be a convenient measure of turbulence. Analyses at the National Advisory Committee on Aeronautics [129] and by the American Airlines System [130] indicate that an aircraft equipped to detect turbulence by this means could avoid the most serious gusts, with a substantial improvement in aircraft safety and passenger comfort. Studies of the design of a 5.7-cm radar intended specifically for this purpose are now in progress under the auspices of Aeronautical Radio, Inc.

Jones and the NACA also encountered considerable turbulence in cumulus clouds which gave no echo at all (because the particles were undetectably small). A radar study of such turbulence and that in clear air is probably possible only by observing the motion of "window" (freely falling strips of light metal foil) distributed by aircraft. This technique has been tried successfully by Warner and Bowen [121].

3.6. Quantitative Observations

The quantitative exploitation of the power of the weather echo is not easy; but useful and reliable measurements of rain intensity have shown that it is worthwhile. Such work involves the theoretical equation (1). Attempts to verify this equation have been made in two stages.

The relatively easy, and for practical purposes essential, step was the demonstration in 1947 by Marshall, Langille, and Palmer [86] that the power received from rain on 10-cm equipment was proportional to $Z = \Sigma D^6$. This finding is in agreement with the Rayleigh approximation and is valid approximately at all wavelengths (see Fig. 24). Using equation (3), the above proportionality may be written in the form

$$\overline{P_r} = AR^{1.6}/r^2 \tag{8}$$

which is valid in the absence of attenuation, and provided the cross-section of the beam is filled with target rain. The calibration constant A, which depends on the characteristics of the radar and the dielectric factor, can be determined by using a rain gage. Equation (8) (or a variant if a Z/R correlation differing from equation (3) is preferred), is the sufficient basis for quantitative studies of precipitation structure (Figs. 10 and 11) or for precipitation measurements.

The earliest operational checks of 3- and 10-cm radars against a network of rain gages was probably that undertaken in Ohio, in 1948, in conjunction with the Thunderstorm Project (Bunting and Latour [122]). On a larger scale, similar observations have been carried on by the Illinois State Water Survey. Stout [123] reported a series of radar observations over several watersheds on which fairly dense networks of automatic rain gages had been arranged. Stout and Neill [124] described in some detail their experience on one such network, 70 square miles in area with 33 gages. At single points they found the disagreement between radar and gage indications disappointing, but when they plotted rainfall contours, on the basis of their two sets of observations, agreement was generally satisfactory. They expressed the accuracy of their radar results in terms of the density of an equivalent network of rain gages. This equivalent density, which varied in their observations from one gage per 3 square miles to one gage per 200 square miles, appeared to be a sensitive function of the intensity of precipitation [125], but in any case it was much greater than is available in routine hydrological observations. A good part of the Illinois radar measurements was made on 3-cm equipment and to some extent suffered from attenuation, no correction for which was attempted.

All the observers mentioned thus far measured echo powers by reading heights on a calibrated A-scope, or obtained echo power contours by varying the receiver gain in known steps until the signal just disappeared. The pulse integrator as used by Austin and Richardson [126] provided a much more accurate, although slower, method. These authors measured echo powers corresponding to areas 10 miles deep and effectively 3 degrees wide at ranges varying between 12 and 62 miles from the radar.

They thus sacrificed resolution but felt that the loss in accuracy involved was not serious "since the limits of accuracy are inherently broad because of variations in the drop size distribution." To increase accuracy somewhat, Austin and Richardson suggested the use of three different calibration constants depending on the type of precipitation studied. Although the 10 miles mentioned may be an unnecessarily long distance for averaging, there would seem to be general agreement as to the wisdom of measuring areal averages. Radar does not give reliable spot values for R at a specified point on the ground, but it will readily and reliably supply rainfall values averaged over areas large compared to a contributing region. In a matter of minutes, the radar can thus draw a rainfall map for an area of several thousand square miles: it is not an instrument suitable for replacing, but rather for supplementing, rain gages, although in many cases the information traditionally sought by means of gages can be more efficiently obtained by radar.

Snow measurements by radar have been attempted by Langille and Thain [104]. Their experiment verified the scattering theory for snow as had previously been done for rain [86]. Their results (for snow) were later shown [103] to fit the rain equation (8), provided that R is taken to be the rate of precipitation in millimeters of melted water per hour. On one occasion, Langille and Thain were able to make an *absolute* check of the radar theory. They measured directly all the known factors entering into equation (1) and in this way obtained a value about 0.6 (−4 db) for the factor F. As has been pointed out before, the reason for this term is not clearly understood.

Absolute verification of the radar weather equation is a complicated procedure. Hooper and Kippax [127] were probably the first to attempt it both for rain and snow, and they obtained complete agreement of theory and measurements on the three wavelengths, 9.1, 3.2 and 1.25 cm. Similar measurements were made by Hood [89] on rain. Like Hooper and Kippax, he measured echo power on an A-scope calibrated by means of a signal generator and an attenuator. Unlike them, his echo powers were always lower than expected by theory; at 10 cm wavelength the discrepancy was −4.5 db, at 3 cm it was −10 db. These results may be expressed with reference to equation (1) by giving to the empirical factor F the values 0.37 or 0.1, respectively.

The accuracy in the above experiments is somewhat open to question because of the great difficulty of absolute power measurements on an A-scope. It must be remembered that the precipitation echo continually fluctuates so that the A-scope trace has no easily recognized mean. Comparison with a steady signal from the signal generator is therefore always subjective. Photographing several A-scope traces is not helpful either, unless the number of traces is very large and careful account is taken of the statistics involved [114]. These difficulties were overcome by Austin and Williams [80] in their careful and systematic check of radar weather theory by the use of the pulse integrator, the output of which is a meter reading representing the average of a large number of echo amplitudes. The response of their pulse integrator was constantly checked against an attenuated signal generator, while the performance of the whole radar system was calibrated against a balloon-borne standard target. The effect of the shape of the beam on echoes of extended targets, previously not considered, was also taken into account. In spite of these precautions, Austin and Williams obtained echo powers from rain some 7 db smaller than had been expected on theoretical grounds. It is on account of this finding, for which no complete explanation is available, that the factor F has been inserted into equation (1) with a suggested value of 0.2.

The only quantitative observations of fogs to date are those made in 1953 by investigators of the Cloud Physics Section of the Air Force Cambridge Research Center. According to a preliminary report [128], liquid-water content and drop-size distributions of fogs were measured on the Atlantic coast, while radar echo powers were measured after the manner of Austin and Williams. Again less power was received than theory had indicated; a difference of 10 db ($F = 0.1$) was recorded. This discrepancy, too, is unresolved at present, although it was suggested that there might possibly have been a considerable difference in the droplet size distribution over the sea (in the contributing region) and on the shore where the samples were taken.

3.7. Summary

The accidental good fortune which makes these radar measurements possible is the applicability of the approximate Rayleigh theory in microwave scattering from rain, snow, and cloud. Echo power can be related to $Z = \Sigma D^6$, the sum of the sixth powers of the particle diameters in unit volume. Usable correlations between the radar Z and the rate of precipitation, or the cloud density, have been developed; their quality is still subject to improvement, as are the particle distribution data on which they depend.

The interesting fact that radar signals are weaker than calculated from theory has not been fully explained, but this has been no obstacle to practically useful measurements of the rate of rainfall. The power of radar to produce reliable rainfall maps for thousands of square miles in a matter of minutes has been demonstrated. Snow measurements on a routine basis have not yet been tried, nor are measurements of cloud densities by radar far enough advanced to indicate with certainty what fraction of clouds may be observed.

The fluctuating nature of the weather echo appreciably slows down quantitative observations as it necessitates the averaging of many independent signals from one region. At the same time there is some hope that measurements of the rate of the fluctuations can be made to serve in studies of atmospheric turbulence.

Although the study of precipitation pattern by radar has made enormous strides without drawing heavily on the detailed theory of the radar echo, quantitative observations of pattern which make full use of this theory may be expected to contribute increasingly, and possibly decisively, to our understanding of the mechanism of precipitation. The additional information in contoured pictures (e.g., Figs. 10 (lower half), 11, or 12) as compared to more qualitative records (Figs. 10 (upper half) or 5), is impressive.

Acknowledgments

The authors are members of the Stormy Weather Group at McGill University. Much of the work described in this article derives from the research work of this group which is sponsored jointly by the United States Air Force, Contract AF 19(122)-217 and the Defence Research Board of Canada, Project D-48-95-11-08.

List of Symbols

- A calibration constant of weather radar
- A_e effective area of radar antenna for extended targets
- c cloud attenuation (db/mi, one way)
- D drop diameter
- d density of snowflake (total volume of flake/mass)
- F uncertainty factor. Introduced into equation (1) to indicate the disagreement between measured and calculated values of echo power. A discussion of values of F deduced from various observations is given at end of section 3.6.
- g gas attenuation (db/mi, one way)
- h length of radar pulse
- K constant for rain attenuation (db/mi (mm/hr)$^{-\alpha}$, one way)
- K' constant for cloud attenuation (db/mi (gm/m^3)$^{-1}$, one way)
- k total attenuation factor
- M cloud density (gm/m^3)
- m complex refractive index
- P_t power transmitted by radar during transmission period
- $\overline{P_r}$ average value of echo power received at radar
- p rain attenuation (db/mi, one way)
- R rate of precipitation (mm of rain water or melted snow per hour)
- r range from radar (usually in miles)
- r_0 range from radar up to which cross-section of beam is filled in one or both dimensions
- T time required for power of weather echo to reach an effectively independent value

t time
v terminal velocity of particle
W horizontal velocity of radar pattern
w component of wind in x-direction
x horizontal distance
$Z = \Sigma D^6$, the sum of the sixth powers of particle diameters in unit volume
z vertical distance (measured downwards) (section 2.2)
$\alpha = \pi D/\lambda$ parameter in Mie's and Rayleigh's theory of scattering
α power of R occurring in expression for rain attenuation
ϵ ratio of exact to approximate back scatter cross-section of spheres
κ dielectric factor
λ radar wavelength
ρ ratio of axis to diameter of spheroid (ρ_0 oblate spheroid, ρ_p prolate spheroid)
$\rho(\tau)$ autocorrelation function
σ back scatter cross-section (dimensions of area)
$\Sigma\sigma$ reflectivity, the sum of the back scatter cross-sections of the particles in unit volume
τ interval in time for autocorrelation study

References

1. Wexler, H. (1945). Introduction to [9].
2. Ryde, J. W. (1947). The attenuation and radar echoes produced at centimetre wavelengths by various meteorological phenomena. Meteorological Factors in Radio Wave Propagation. The Physical Society (London).
3. Lord Rayleigh (1871). On the scattering of light by small particles. *Phil. Mag.* **41,** 447.
4. Mie, G. (1908). Beiträge zur Optik trüber Medien, speziell kolloidaler Metallösungen. *Ann. phys.* [4] **25,** 377.
5. Burrows, C. R., and Attwood, S. S., eds. (1949). "Radio Wave Propagation" Academic Press, New York.
6. Siegert, A. J. F. (1943). On the fluctuations in signals returned by many independently moving scatterers. Massachusetts Institute of Technology, Radiation Laboratory, Report 465.
7. Siegert, A. J. F. (1946). Fluctuations in the return signal from random scatterers. Massachusetts Institute of Technology, Radiation Laboratory, Report 773.
8. Kerr, D. E., ed. (1951). *Propagation of Short Radio Waves,* McGraw-Hill, New York, 728 pp.
9. Maynard, R. H. (1945). Radar and weather. *J. Meteorol.* **2,** 214–226.
10. Bent, A. E. (1946). Radar detection of precipitation. *J. Meteorol.* **3,** 78–84.
11. Ligda, M. G. H. (1951). Radar storm observation. *in* Compendium of Meteorology, American Meteorological Society, Boston, pp. 1265–1282.
12. Wexler, R. (1951). Theory and observation of radar storm detection. *in* Compendium of Meteorology, American Meteorological Society, Boston, pp. 1203–1289.
13. Van Bladel, J. (1955). "Les Applications du Radar à l'Astronomie et à la Météorologie." Gauthier-Villars, Paris.
14. Proceedings, (1952). Conference on water resources, October 1–3, 1951. Illinois State Water Survey, Urbana, Illinois, 335 pp.
15. Proceedings, (1952). Third Radar Weather Conference, September 15–17, 1952, McGill University, Montreal.

16. Proceedings, (1953). Conference on Radio Meteorology, November 9–12, 1953. The University of Texas, Austin, Texas.
17. Schiff, W. J., and Williams, E. L. (1951). Details of the production model, AN/CPS-9 storm detector radar. Conference Proceedings pp. 223–225; see ref. [14].
18. Byers, H. R., and Coons, R. D. (1947). The bright band in radar cloud echoes and its probable explanation. *J. Meteorol.* **4**, 75–81.
19. Swingle, D. M. (1950). 2800 Mc/s vertical incidence reflections from the troposphere. Harvard University, Cruft Laboratory Technical Report No. 113, 21 pp.
20. Marshall, J. S. (1951). Intensity of radar echoes from snow. Paper read to the American Meteorological Society, Washington, May (unpublished).
21. Bowen, E. G. (1951). Radar observations of rain and their relation to mechanisms of rain formation. *J. Atm. and Terrest. Phys.* **1**, 125–140.
22. Browne, I. C. (1952). Precipitation streaks as a cause of radar upper bands. *Quart. J. Roy. Meteorol. Soc.* **78**, 590–595.
23. Lhermitte, R. (1952). Les "bandes supérieures" dans la structure vertical des échos de pluie. *Compt. rend. acad sci.* **235**, 1414–1416.
24. Marshall, J. S. (1953). Precipitation trajectories and patterns. *J. Meteorol.* **10**, 25–29.
25. Langleben, M. P. (1954). The terminal velocity of snow aggregates. *Quart. J. Roy. Meteorol. Soc.* **80**, 171–181.
26. Gunn, K. L. S., Langleben, M. P., Dennis, A. S., and Power, B. A. (1954). Radar evidence of a generating level for snow. *J. Meteorol.* **11**, 20–26.
27. Cunningham, R. M., and Atlas, D. (1954). Growth of hydrometeors and calculations based on aircraft and radar observations. Proceedings, Toronto Meteorological Conference, September 9–15, 1953, The Royal Meteorological Society, London.
28. Wexler, R., and Austin, P. M. (1953). Radar signal intensity from different levels in continuous precipitation. Conference Proceedings p. VII-3; see ref. [16].
29. Kerker, M., Langleben, P., and Gunn, K. L. S. (1951). Scattering of microwaves by a melting spherical ice particle. *J. Meteorol.* **8**, 424.
30. Labrum, N. R. (1952). Some experiments on centimetre wavelength scattering by small obstacles, *J. Appl. Phys.* **23**, 1320–1323.
31. Austin, P. M., and Bemis, A. C. (1950). A quantitative study of the "bright band" in radar precipitation echoes. *J. Meteorol.* **7**, 145–151.
32. Labrum, N. R. (1952). The scattering of radio waves by meteorological particles. *J. Appl. Phys.* **23**, 1324–1330.
33. Hooper, J. E. N., and Kippax, A. A. (1950). The bright band—A phenomenon associated with radar echoes from falling rain. *Quart. J. Roy. Meteorol. Soc.* **76**, 125–132.
34. Jones, R. F. (1951). Aircraft observations of radar reflecting particles above the freezing level. A paper of the Meteorological Research Committee (London) MRP 683.
35. Atlas, D., and Banks, A. C. (1950). A virtual echo-layer above the bright band. *J. Meteorol.* **7**, 402–403.
36. Wexler, R., and Honig, J. (1953). Atmospheric cooling by melting snow. Conference Proceedings p. VII-5; see ref. [16].
37. Cunningham, R. M. (1952). The distribution and growth of hydrometeors around a deep cyclone. Massachusetts Institute of Technology, Weather Radar Research, Technical Report, No. 18, 59 pp.

38. Byers, H. R., and Braham, R. R. (1949). The Thunderstorm. U.S. Government Printing Office, Washington, D. C., 287 pp.
39. Langille, R. C., and Gunn, K. L. S. (1948). Quantitative analysis of vertical structure in precipitation. *J. Meteorol.* **5,** 301–304.
40. Atlas, D. (1953). Device to permit radar contour mapping of rain intensity in rainstorms. U.S. Patent 2,656,531.
41. Battan, L. J. (1953). Observations on the formation and spread of precipitation in convective clouds. *J. Meteorol.* **10,** 311–324.
42. Workman, E. J., and Reynolds, S. E. (1949). Electrical activity as related to thunderstorm cell growth. *Bull. Am. Meteorol. Soc.* **30,** 142–144.
43. Jones, R. F. (1950). The temperatures at the top of radar echoes associated with various cloud systems. *Quart. J. Roy. Meteorol. Soc.* **76,** 312–330.
44. Hilst, G. R., and MacDowell, G. P. (1950). Radar measurements of the initial growth of thunderstorm precipitation cells. *Bull. Am. Meteorol. Soc.* **31,** 95–99.
45. Smith, E. J. (1951). Observations of rain from non-freezing clouds. *Quart. J. Roy. Meteorol. Soc.* **77,** 33–43.
46. Rigby, E. C., Marshall, J. S., and Hitschfeld, W. (1954). Development during fall of raindrop size distributions, *J. Meteorol.* **11,** 362–372.
47. Bowen, E. G. (1950). The formation of rain by coalescence. *Australian J. Sci. Research.* **3,** 193–213.
48. Reynolds, S. E., and Braham, R. (1952). Significance of the initial radar echo. *Bull. Am. Meteorol. Soc.* **33,** 123.
49. Braham, R. R., Reynolds, S. E., and Harrell, J. H. (1951). Possibilities for cloud seeding as determined by a study of cloud height versus precipitation. *J. Meteorol.* **8,** 416–418.
50. Dennis, A. S. (1954). Initiation of showers in cumuli by snow. *J. Meteorol.* **11,** 157–162.
51. Wexler, R. (1947). Radar detection of a frontal storm, June 18, 1946. *J. Meteorol.* **4,** 38–44.
52. Wexler, R. (1947). Radar photographs of a frontal wave. *J. Meteorol.* **4,** 69–71.
53. Austin, P. M. (1951). Radar observations of a frontal storm. *Bull. Am. Meteorol. Soc.* **32,** 136–145.
54. Stout, G. E. (1953). Radar in single station analysis. Conference Proceedings p. XI-2; see ref. [16].
55. Swingle, D. M., and Richards, W. J. (1953). The variability of the motion of radar precipitation lines. Conference Proceedings p. XI-4; see ref. [16].
56. Swingle, D. M., and Rosenberg, L. (1953). Mesometeorological analysis of cold front passage using radar weather data. Conference Proceedings p. XI-5; see ref. [16].
57. Ligda, M. G. H. (1952). The horizontal motion of small precipitation areas. Conference Proceedings p. D41–48; see ref. [15].
58. Bunting, D. C., Gentry, R. C., Latour, M. H., and Norton, G. (1951). Florida hurricanes of 1950. *Eng. Prog. Univ. Florida* **5,** 36 pp.
59. Jorgensen, R. C., and Gerdes, W. F. (1951). The Dow Chemical Company leads industry in using radar for hurricane detection. *Bull. Am. Meteorol. Soc.* **32,** 221–224.
60. Stout, G. E., and Huff, F. A. (1953). Radar records Illinois tornado-genesis. *Bull. Am. Meteorol. Soc.* **34,** 281–284.
61. Freeman, J. C. (1954). Pictures of recent Texas tornado. A paper delivered at the New York meeting of the American Meteorological Society, January.

62. Jones, H. L. (1953). Correlation between sferics and tornado incidence. Conference Proceedings p. V-1; see ref. [16].
63. Ligda, M. G. H. (1950). Lightning detection by radar. *Bull. Am. Meteorol. Soc.* **31,** 279–283.
64. Browne, I. C. (1951). A radar echo from lightning. *Nature* **167,** 438.
65. Marshall, J. S. (1953). Frontal precipitation and lightning observed by radar. *Can. J. Phys.* **31,** 194–203.
66. Miles, V. G. (1953). Radar echoes associated with lightning. *J. Atm. and Terrest. Phys.* **3,** 258–261.
67. Hewitt, F. J. (1953). The study of lightning streamers with 50 cm radar. *Proc. Phys. Soc. (London)* **B66,** 895.
68. Swingle, D. M. (1953). Thunderstorm identification on the PPI. Conference Proceedings p. V-6; see ref. [16].
69. Gould, W. B. (1951). Cloud detection by radar. Conference Proceedings pp. 209–219; see ref. [14].
70. Fisher, F. W., Gould, W. B., and Greenslit, C. L. (1952). Radar cloud base and top indicator. Conference Proceedings p. E-1; see ref. [15].
71. Swingle, D. M. (1952). An operational comparison of radar cloud detection at 1.25 cm and 0.86 cm. Conference Proceedings p. B-21; see ref. [15].
72. Swingle, D. M. (1953). A review of results of operational test of radar cloud detection equipment. Conference Proceedings p. X-4; see ref. [16].
73. Plank, V. G., Atlas, D., and Paulsen, W. H. (1952). A preliminary survey of cloud and precipitation detection. Conference Proceedings p. B-13; see ref. [15].
74. Gould, W. B. (1952). Recent observations on radar reflections from the lower atmosphere. Conference Proceedings p. F-25; see ref. [15].
75. Crain, C. M., Deam, A. P., and Gerhardt, J. R. (1953). Measurement of index-of-refraction fluctuations and profiles. *Proc. I.R.E.* **41,** 289–290.
76. Browne, I. C. (1953). Radar echoes at vertical incidence from a horizontally stratified atmosphere. *Quart. J. Roy. Meteorol. Soc.* **79,** 157.
77. Friend, A. W. (1949). Theory and practice of tropospheric sounding by radar. *Proc. I.R.E.* **37,** 116–138.
78. Plank, V. G. (1953). A meteorological study of angels at 1.25 cm. Conference Proceedings p. IV-3; see ref. [16].
79. Atlas, D., Paulsen, W. H., Donaldson, R. J., Chmela, A. C., and Plank, V. G. (1953). Observations of the sea breeze by 1.25 cm radar. Conference Proceedings p. XI-6; see ref. [16].
80. Austin, P. M., and Williams, E. L. (1951). Comparison of radar signal intensity with precipitation rate. Massachusetts Institute of Technology, Weather Radar Research, Technical Report, No. 14, 43 pp.
81. Lowan, A. (1949). Tables of Scattering Functions for Spherical Particles. U.S. Government Printing Office, Washington, D. C.
82. Aden, A. L., and Kerker, M. (1951). Scattering of electromagnetic waves from two concentric spheres. *J. Appl. Phys.* **22,** 1242–1246.
83. Langleben, M. P., and Gunn, K. L. S. (1952). Scattering and absorption of microwaves by a melting ice sphere. McGill University, Stormy Weather Group, Scientific Report MW-5, 33 pp.
84. Laws, J. O., and Parsons, D. A. (1943). The relation of raindrop-size to intensity. *Trans. Am. Geophys. Un.* **24,** part II, 452.
85. Haddock, F. T. (1948). Scattering and attenuation of microwave radiation through rain. Naval Research Laboratory, Washington, D. C. (unpublished manuscript).

Institute of Technology, Weather Radar Research, Technical
22A,B,C.
86. Bartnoff, S., Paulsen, W. H., and Atlas, D. (1952). Experimental s
cloud and rain echoes. Conference Proceedings p. G-1; see ref. [15].
87. Williams, E. L. (1949). The pulse integrator: Part A, Description of th
ment and its circuitry. Massachusetts Institute of Technology, Weather
Research, Technical Report, No. 8A, 35 pp.
88. Ligda, M. G. H. (1951). The pulse integrator: Part B, Methods of measuring
power of radar precipitation echoes. Massachusetts Institute of Technolog
Weather Radar Research, Technical Report, No. 8B.
Marshall, J. S., and Hitschfeld, W. (1953). The interpretation of the fluctuating
echo from randomly distributed scatterers, Part I, *Can. J. Phys.* **31**, 962–994.
Wallace, P. R. (1953). The interpretation of the fluctuating echo from randomly
distributed scatterers, Part II, *Can. J. Phys.* **31**, 995–1009.
Austin, Pauline, (1952). Observations of the amplitude distribution function for
radar echoes from precipitation. Conference Proceedings p. G-19; see ref. [15].
Hilst, G. R. (1949). Analysis of the audio-frequency fluctuations in radar storm
echoes: A key to the relative velocities of the precipitation particles. Massachusetts Institute of Technology, Weather Radar Research, Technical Report,
No. 9A, 60 pp.
Williams, E. L. (1950). Radar signal spectrograph (Rasaph): Description of the
instrument and its circuitry. Massachusetts Institute of Technology, Weather
Radar Research, Technical Report, No. 9B, 24 pp.
Fleisher, A. (1951). The use and limitations of Rasaph. Conference Proceedings
p. 263–267; see ref. [14].
…nes, R. F. (1952). Turbulence in relation to the radar echoes from cumulo-
…mbus clouds. Conference Proceedings p. A-37; see ref. [15]. (Amplified in
…itish Meteorological Office Professional Note 109, in press.)
…arner, J., and Bowen, E. G. (1953). A new method of studying the fine struc-
…re of air movements in the free atmosphere. *Tellus* **5**, 36–41.
…nting, D. C., and Latour, M. H. (1951). Radar-rainfall studies in Ohio, *Bull.*
…n. Meteorol. Soc. **32**, 289–294.
…out, G. (1952). Rain gage network observations. Conference Proceedings
… D-1; see ref. [15].
…ut, G. E., and Neill, J. C. (1953). Utility of radar in measuring areal rainfall.
…ll. Am. Meteorol. Soc. **34**, 21–27.
…ill, J. C. (1953). Variance of areal-mean thunderstorm rainfall estimates.
…nference Proceedings p. IX-2; see ref. [16].
…tin, P. M., and Richardson, C. (1952). A method of measuring rainfall over
…area by radar. Conference Proceedings p. D-13; see ref. [15].
…per, J. E. N., and Kippax, A. A. (1950). Radar echoes from meteorological
…ipitation. *Proc. Inst. of Elec. Eng.* (London), *Part I.* **97**, 89–97.
…aldson, R. J., Atlas, D., Paulsen, W. H., Cunningham, R. M., and Chmela,
…C. (1953). Quantitative 1.25-cm observations of rain and fog. Conference
…eedings p. VII-6; see ref. [16].
…fson, H. B. (1953). Some possible reductions in gust loads through the use of
…r in transport airplanes. *Bull. Am. Meteorol. Soc.* **34**, 187–191.
…, R. W., White, F. C., and Armstrong, L. W. (1949). The development of an
…rne radar method of avoiding severe turbulence and heavy precipitation in
…recipitation area of thunderstorms and line squalls. Final report on Task
…1 (Contract No. a(s)-9006, Bureau of Aeronautics), American Airlines
…m, Operational Development Branch.

Marshall, J. S., Langille, R. C., and Palmer, W. McK. (19
of rainfall by radar. *J. Meteorol.* **4,** 186–192.
Wexler, R. (1948). Rain intensities by radar. *J. Meteorol*
. Marshall, J. S., and Palmer, W. McK. (1948). The distrib
size. *J. Meteorol.* **5,** 165–166.
89. Hood, A. D. (1950). Quantitative measurements at 3 a
intensities from precipitation. National Research Cou
No., **2155,** 15 pp.
90. Best, A. C. (1950). The size distribution of raindrops.
Soc. **76,** 16–36.
91. Twomey, S. (1953). On the measurement of precipitati
10, 66–67.
92. Blanchard, D. C. (1953). Raindrop size distributio
Meteorol. **10,** 457–473.
93. Bowen, E. G., and Davidson, K. A. (1951). A raindro
Roy. Meteorol. Soc. **77,** 445–449.
94. Mason, B. J., and Ramanadham, R. (1953). A photoe
eter. *Quart. J. Roy. Meteorol. Soc.* **79,** 490–495.
95. Adderley, E. E. (1953). The growth of raindrops in clo
Soc. **79,** 380–388.
96. Jones, D. M. A., and Dean, L. A. (1953). Some obs
of falling raindrops. Conference Proceedings p. VIII
97. Magono, C. (1954). On the shape of water drops
Meteorol. **11,** 77–79.
98. Atlas, D., and Boucher, R. J. (1952). The relation
size distribution. Conference Proceedings p. B-1; se
99. Atlas, D., Kerker, M., and Hitschfeld, W. (1953). S
non-spherical atmospheric particles. *J. Atm. and T*
100. Gans, R. (1912). Über die Form ultramikroskopisc
37, 881–900.
101. Browne, I. C., and Robinson, N. P. (1952). Cross-
ing band. *Nature* **170,** 1078.
102. Hunter, I. M. (1954). Polarization of radar echoes
tion. *Nature* **173,** 165–166.
103. Marshall, J. S., and Gunn, K. L. S. (1952). Measu
radar. *J. Meteorol.* **9,** 322–327.
104. Langille, R. C., and Thain, R. S. (1951). Some
three-centimeter radar echoes from falling snow.
105. Atlas, D., and Banks, H. C. (1951). The interpr
from rainfall. *J. Meteorol.* **8,** 271–282.
106. Hitschfeld, W., and Bordan, J. (1954). Errors in
of rainfall at attenuating wavelengths. *J. Mete*
107. Hitschfeld, W., and Marshall, J. S. (1954). Effe
wavelength for weather detection by radar. *Pr*
107a. Gunn, K. L. S., and East, T. W. R. (1954). M
tion particles. *Quart. J. Roy. Meteorol. Soc.* **80,**
108. Lord Rayleigh (1945 Reprint). The Theory of S
New York, 480 pp. (pp. 35–42).
109. Fleisher, A. (1953). The radar as a turbulenc
ceedings p. XI-7; see ref. [16].
110. Fleisher, A. (1953, 1954). Information contain

111.
112.
113.
114.
115.
116.
117.
118.
119.
120. J
121. W
122. B
123. St
124. St
125. Ne
126. Au
127. Ho
128. Do
129. Tol
130. Aye

Methods of Objective Weather Forecasting

IRVING I. GRINGORTEN

Air Force Cambridge Research Center, Cambridge, Mass.

1. Introduction

1.1. Definition

An objective forecast is one that is made without recourse to the personal judgment of the forecaster. Strictly speaking, if two forecasters were given copies of one manual describing a forecast method and placed in separate rooms with current data, they would make identical forecasts.

1.2. Purpose

The desirability of objectivity should be apparent to forecasters and forecast users alike. Its primary purpose is to bring forecasters closer to that elusive goal: accuracy! In an objective method, meteorological parameters appear in formulas or relations that are theoretical or empirical, in graphical, tabular, or algebraic forms. Not only should the objective device have, on paper, the knowledge and techniques by which an experienced forecaster makes his decision, but it should surpass the limited mental process of even the best forecasters in coordinating previous meteorological events into a unified pattern relating cause to effect.

After a few weeks of forecasting, however, with or without an objective system, any meteorologist will envy the astronomer's science [1] and predictions that are never for a moment doubted by the public at large. The weather forecaster, contemplating the low degree of accuracy of his product, might settle for a slight improvement and accept perfect forecasting as utopia.

But there are other reasons justifying objectivity. Because of the present subjective nature of forecasting it is mainly the forecasters of long experience who have enjoyed the confidence of the operating public. Objective methods or devices become tools for the less experienced forecaster. They attempt to compensate for his undeveloped intuition or an unacquired ability to remember details and to coordinate them in his mind. An objective device can concentrate a forecaster's attention upon the parameters that have worked best as predictors of the weather at his station. It can give him the best available statistics or the best interpretation of the statistics, or provide the means for rapidly referring to an historical set of weather charts to find situations similar to the problem situation.

Another goal of objectivity, which would be difficult for the subjective forecaster to achieve, is to determine quantitatively the probability of a future event following a given meteorological condition. To illustrate this advantage let us suppose that the forecast is concerned with the onset of fog at a sea-coast airport. With a prevailing onshore wind the experienced forecaster will anticipate fog, try to evaluate other conditions concurrent with the prevailing wind, and then predict fog if he is convinced that it will develop. Frequently he will be right, but sometimes wrong. Using an objective system specially designed for the job, the forecaster could state the statistical frequency or probability of fog following the observed antecedents, or his probable degree of accuracy.

1.3. Interest

The interest in objective forecasting has existed for many years [2], perhaps waning at times only to develop again. Figure 1 gives the numerical count, by years, of articles in Kramer's bibliography [3] on objective forecasting. It may be no casual coincidence that papers on objectivity began to appear after Lord Kelvin's famous remark (too famous to quote here) that relates our comprehension of a subject to our ability

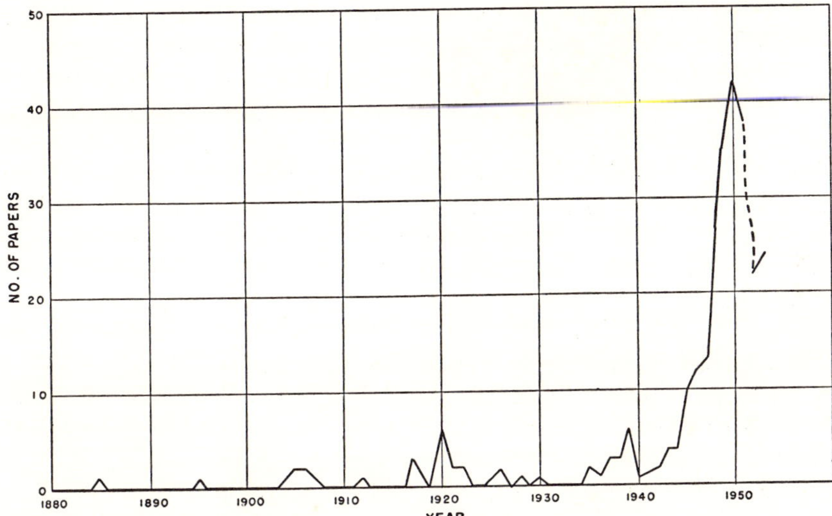

FIG. 1. Number of papers on objective forecasting reviewed in Meteorological Abstracts and Bibliography, Vol. 3, No. 5, American Meteorological Society.

to measure its parameters numerically. The U. S. Weather Bureau, through its medium the Monthly Weather Review, has been publishing objective studies since 1906 [4]. But the interest in objectivity need not be judged solely by the published word. Many forecasting offices can be visited where the forecasters have been using devices specially designed for use by their station personnel to answer several key operational questions at the stations.

1.4. Limitations

The limitations and difficulties of objective forecasting are painful. Atmospheric motions and conditions are far too turbulent and complex to be reduced to a single mathematical model or to be unified by a system of equations completely relating cause and effect. The rapid electronic computers of recent years might cope with this problem if the simplifying assumptions, that still must be made, will prove admissible or tenable. In recent German publications [5] one author states that the complexity of the atmospheric motions "makes computations a veritable Herculean task and the most rapid electronic brain remains as hopeless as the dumbest Neanderthal man."

By its existence alone an objective system cannot increase the potential accuracy of forecasts. This might be done only by a better knowledge of the physics of the atmosphere and by improved models of the atmosphere. An objective device should be able to squeeze J. J. George's sponges drier [6] and extract the maximum degree of accuracy of which the forecaster, with his present knowledge and experience, is capable.

Most investigators escape some of the complexity of the problem by limiting their investigations to one or two isolated forecast problems, such as "winter rainfall" or "summer thunderstorms" at a given locality or in a well-defined region. But such studies will answer objectively only a handful of questions. If a system is designed to answer the question "rain or no rain," other questions must be answered subjectively, such as visibility in the rain, cloud developments, heights, associated gusts, icing in the clouds, and so on.

As opposed to subjective forecasting, an objective system can use only a limited number of parameters as predictors of future events. Even a detailed and comprehensive system must neglect some very desirable parameters. An objective method does not take into account the spot checking of observations and trends in the problem area. The use of raw observational data, on which the predictors are based, itself implies certain assumptions. For example, the reading of the temperature in a thermometer shelter must be assumed to represent the temperature of the mass of air in its area. Or, the "observed" values of pressure, temperature, and humidity at a given height in the atmosphere imply that the atmosphere was in a state of equilibrium when the measurements were made.

If the objective tool is derived statistically, or based upon the past performance of the weather, its use for the prediction of novel or rare events, such as a tornado in New England, is almost axiomatically suspect. Moreover, the weather may undergo a secular change, enough to invalidate some statistical inferences applied to future events.

It is little wonder that forecasting today is, in large measure, an art. There is heavy reliance upon "qualified" forecasters who have difficulty to submit their techniques to writing so that their skills might be had for the reading.

1.5. Historical Attitudes

For the above reasons, objective forecasting itself has become a problem in meteorology, tackled in the literature and over the conference table. A survey by Allen [7] appeared in 1949, followed by Kramer's excellent bibliography in 1952 [3]. In the Compendium of Meteorology [8] several authors have attacked the problem, directly or indirectly. Willett [9] states, "It is a sad commentary on the scientific status of weather forecasting that even today the methods remain so empirical and so dependent upon experience and subjective interpretation that the development of the best forecasters still requires years of experience and the right temperament and interest. The primary problem remains . . . making it scientifically objective." However, the same author states that by long experience a forecaster usually develops a better sense of the probability of any weather occurrence at his station than he is likely to get from statistics. Gordon Dunn [10] states, ". . . the possibility of forecasting the weather wholly on an objective basis seems very remote," and adds that the objective methods have met with passive resistance from forecasters "who feel that the derived formulas and diagrams oversimplify the problem and that other contributory variables are frequently neglected." Bundgaard [11], however, in writing on short-range forecasting procedures, describes several objective techniques in detail without casting any doubt on their effectiveness, and treats them as accepted pieces in a forecaster's box of tools.

The Compendium of Meteorology has one article devoted completely to objective forecasting [12], in which the authors, Allen and Vernon, claim that, in spite of the difficulties, the methods described in recent literature have enjoyed "astonishing success in producing useful results."

Sutcliffe [13] thinks that the objective statistical methods "may have a big future." Houghton, in giving some personal opinions on the existing status of meteorology [14], said that there was a "possibility" of improvement in forecasting by use of objective aids, ". . . to improve on the individual's skill and memory by using some mechanical procedures." But he also felt that none of the objective methods have been as effective as the subjective methods practiced by an experienced forecaster. "There appears to be no immediate prospect" said Houghton, "of an objective method of forecasting based entirely on sound physical principles."

With such conflicting opinions on the efficacy and applicability of objective forecast methods, the U. S. Air Weather Service, in 1952, invited several civilian and military meteorologists to a conference to answer a number of prepared questions on objectivity and to make recommendations on its future use [15]. The experts generally agreed that there is a distinction between an *objective method* that provides impersonal answers to the forecast problems and an *objective aid* that helps the forecaster to make his personal decisions [12], but they disagreed on whether a forecaster should use one or the other. Some authorities expressed dissatisfaction over the very words "objective" and "subjective." They felt that the topic under discussion was *systematic* procedures of forecasting, to systematize both the knowledge and the thinking of a forecaster, and to present the forecast in a form best suited to the user [16]. The question was asked whether a program of verification would not settle the debate on the use of a forecast method, but this question proved only that there has been no universally accepted scheme of verification.

Claims made at the above-mentioned conference differed with regard to the success of objective methods at military weather stations. This conflict of opinion indicated that the objective procedures could succeed in one location for one or two problems and not succeed at another location or for other problems. A suggestion was made that objective studies relate weather phenomena to synoptic patterns so that the greatest use could be made of prognostic charts when, and if, they are perfected or greatly improved by numerical prediction methods.

1.6. Compromise

From the views expressed in the literature, in conferences, at weather stations by forecasters themselves, and from my own point of view, it seems advisable to extend the scope of this review to include those studies and methods of forecasting which compromise with subjectivity in such attributes as weather-map analysis or in the decision of the forecaster. Some of the forecast studies listed in the references below are completely objective, both in obtaining the antecedent parameters and in the techniques. But many of the papers permit the personal judgment of the forecaster to enter into the forecast. Or they permit the element of subjectivity to enter into the determination of the initial conditions.

On one item, however, this report favors no compromise: the predicted events are defined clearly and explicitly in the references below.

2. Classification of Objective Studies

Recently, Douglas [17] wrote, "It is impossible to give a detailed history of forecasting techniques since they are inherently indescribable." In the face of this categorical statement I shall nevertheless attempt to categorize and describe objective methods of forecasting and let the reader decide how well we can approach this "impossible" task.

Each objective study can be described by three attributes, hence a threefold classification: *predictand*, *predictors*, and *device*.

The *predictand* is that which is to be predicted, such as rainfall. It need not necessarily be a meteorological quantity or condition. For example, in a recent Russian paper, the predictand is "the beginning of the herring spawning off the western coast of South Sakhalin" [18].

The *predictors* are those parameters or conditions of the weather that are used in preparing the forecast.

The *device* is the procedure by which the investigator relates the predictand to the predictors.

In the descriptions that follow, references are mostly to the papers written in 1952 and 1953, assuming that papers prior to 1952 have received adequate treatment by Kramer [3]. I have tried to be influenced neither by an author's claim that he was objective nor by his disclaim of objectivity.

2.1. Devices

It is usual to classify the forecast methods according to whether they have a physical or a statistical basis. Enlarging upon this basic grouping, we obtain the following methods or devices.

2.1.1. Device I. Algebraic Formulas Derived from Physical Principles. In this case the predictands are considered as dependent variables and the predictors as independent variables [19, 20, 21]. The relations may be presented in graphical form. A good example of this technique is Brunt's formula for radiational cooling [22, 23], which presumably is independent of past series of observations and will not be altered with the acquisition of data at any station. Symbolically

$$y = f(x_1, x_2, \ldots, x_n) \tag{1}$$

where y is the predictand, and $x_1, x_2, \ldots, x_n$ are the predictors. This treatment virtually requires that the variables be continuous. In the last two years it has been the tool of authors who *advocate* a forecast method or await the improvement of prognostic charts [24, 25].

2.1.2. Device II. Algebraic Formulas Whose Constants Are Determined Empirically. In this case the data of the station are used to determine the constants. A formula or graph may be based on physical principles or may be a successful empirical relation [26, 27, 28]. Symbolically,

$$y = f(x_1, x_2, \ldots, x_n; a_1, a_2, \ldots, a_m) \tag{2}$$

where $a_1, a_2, \ldots, a_m$ are experimentally determined constants or parameters of the station. The extensive literature on frost and minimum temperatures [29] indicates a universal use of this device for many years.

The U. S. Air Weather Service, in its forecast manual on aircraft icing in clouds, empirically relates the severity of icing to the liquid water content of the cloud, having empirically related the latter to two predictors: temperature at the base of the cloud and height of the flight level above the base of the cloud [30].

2.1.3. Device III. Linear or Curvilinear Regression [31]. The device includes *multiple regression* [32] or extrapolation by periodicity. The method that assumes linearity of the relation between predictors and predictand, at times appearing under the names autocorrelation and cross-correlation [33, 34], is a traditional tool of many inexact sciences. Symbolically,

$$y = b_0 + a_1x_1 + a_2x_2 + \ . \ . \ . + a_nx_n \tag{3}$$

for linear regression or, for a second-order regression,

$$y = b_0 + a_1x_1 + a_2x_2 + c_1x_1^2 + c_2x_2^2 + c_{12}x_1x_2 + \ . \ . \ . \tag{4}$$

The predictand could be the probability of an event ($y = p$), although the difficulty arises that y is not restrained to lie between zero and unity. To dodge this difficulty I had occasion [35] to use an equation such as

$$y = -K_1 \log K_2 \log (1/p) \tag{5}$$

Whereas y was determined by a regression equation such as

$$y = a_1x_1 + a_2x_2 \tag{6}$$

and could vary from $-\infty$ to $+\infty$, p from equation (5) could vary only from zero to unity. For $K_1 = K_2 = 0.798$ the value of p closely approximates the value of y within the range 0.2 to 0.7.

2.1.4. Device IV. Graphical System of Stratification; Use of Scatter Diagrams and Fitting by Eye. This has been a successful technique widely used in the United States and recently in Japan [36, 37, 38, 39]. Symbolically,

$$\begin{array}{lll} F(x_1, x_2) \rightarrow Y_1 & & \\ & F(Y_1, Y_2) \rightarrow Z_1 & \\ F(x_3, x_4) \rightarrow Y_2 & & F(Z_1, Z_2) \rightarrow W \\ F(x_5, x_6) \rightarrow Y_3 & & \\ & F(Y_3, Y_4) \rightarrow Z_1 & \\ F(x_7, x_8) \rightarrow Y_4 & & \end{array} \tag{7}$$

where the x's are the predictors, the Y's and Z's are intermediate quantities, and W is the predictand. Each relation, such as $F(x_1, x_2) \rightarrow Y_1$ is graphically determined, as in Fig. 2, in which observed x_1-values are plotted against simultaneously observed x_2-values. The curved lines of constant Y_1-values are "fitted by eye," the plotted Y_1-values usually being observed values of the predictand. Recently, for a study of precipitation, in the drawing of the family of curves on each scatter diagram, Teweles and Forst [40] paid "attention to a large amount of detail" including precipitation amounts during one 12-hour period, two 6-hour

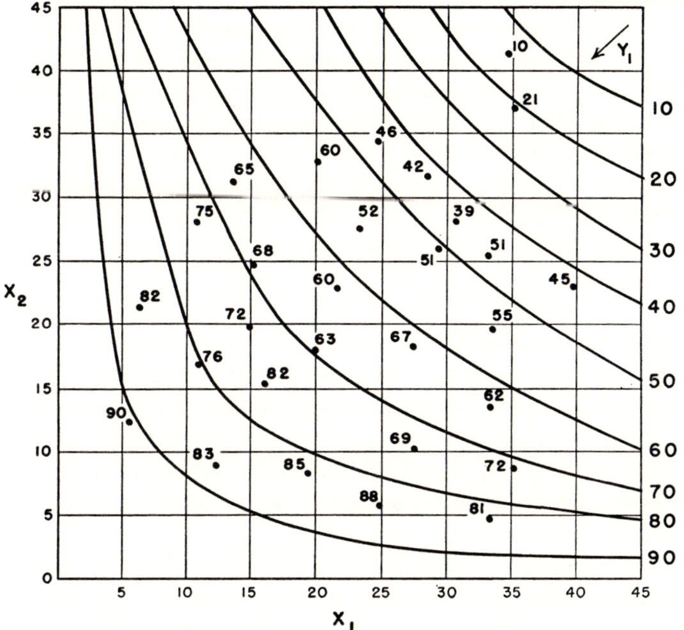

FIG. 2. A scatter diagram in which simultaneous values of two variables (x_1, x_2) are plotted and labeled with the value of a third variable. The family of curves (Y_1 = 10, 20, . . .) are fitted by eye.

periods, and within 50 miles of the station. The "fitting by eye" might prevail throughout the entire procedure until the quantity W is obtained [41].

Between devices III and IV there could be several compromise methods that are based neither completely on the methods of linear correlation nor completely on fitting by eye. But fitting by eye might be used as a first approximation from which a second approximation would be made objectively by a prescribed smoothing method [42].

2.1.5. Device V. Association of Predictors and Predictands by Contingency Tables or Tabulations of Historical Data. This device has been used to avoid the need for continuous variables. Both the predictors and predictands can be defined simply as attributes of the weather, each predictor and predictand being divided into mutually exclusive classes

[43, 44, 45, 46, 47] (e.g., Tables II and III). In some studies this technique is followed by another technique, mostly device IV [48, 49, 50].

In the notation of Yule and Kendall [51], $A, B, C, \ldots$ denote the predictor attributes; $A_1, A_2, A_3, \ldots$ are mutually exclusive classes of the attribute A, exhausting all possibilities of this attribute; $B_1, B_2, B_3, \ldots$ are classes of B, etc.; $X_1, X_2, X_3, \ldots$ are mutually exclusive classes of the predictand, exhausting all possible events defined by the predictand X; $A_qB_rC_s \cdots$ denotes a sub-class of days on which the indicated combination of antecedent conditions prevailed; $(A_qB_rC_s \cdots)$ denotes the number of $A_qB_rC_s \cdots$'s in the period of years that have been studied; and $(A_qB_rC_s \cdots X_k)$ denotes the number of $A_qB_rC_s \cdots$'s that are characterized further by the subsequent event X_k.

The symbolic ratio

$$\frac{(A_qB_rC_s \cdots X_k)}{(A_qB_rC_s \cdots)} \tag{8}$$

constitutes the backbone of the technique, as it gives the observed relative frequency that the event X_k follows the given set of antecedent conditions. The observed relative frequency either is accepted as the probability of the subsequence or is used to establish the limits of the probability [46, 52]. Figure 3 gives the lower 95% confidence limits for observed frequencies in samples that range in size from 1 to 1,000 cases; Fig. 4 gives the upper 95% confidence limits.

The big argument for this technique, in addition to the advantage that the attributes $A, B, C, \ldots; X$ do not need to be continuous variables, is its independence of the true relations between the attributes. It is not necessary to know (or to assume) *how* the predictors are related to the predictands or how they are related to each other. The chief difficulty of the technique is that a class or sub-class of combined predictors will have to be well represented in the historical record before any information of forecast value can be obtained. But many meteorological records are short; 40 years of record for one station are rare. If more predictors are used to define a day's condition there will be fewer days in the historical record that will have been characterized by the same set of conditions. And the smaller the sample of days out of the past, the less reliable will be the observed frequency of an event. The difficulty, however, seems to be of little concern to the forecaster who relies upon an "analogue" method of forecasting, in which the forecaster continues to sub-classify the historical cases until he is left with only one case resembling the existing weather.

The contingency tables are similar to Baur's [43] multiple-correlation tables (Table I). Each cell of a table might give the number of days on

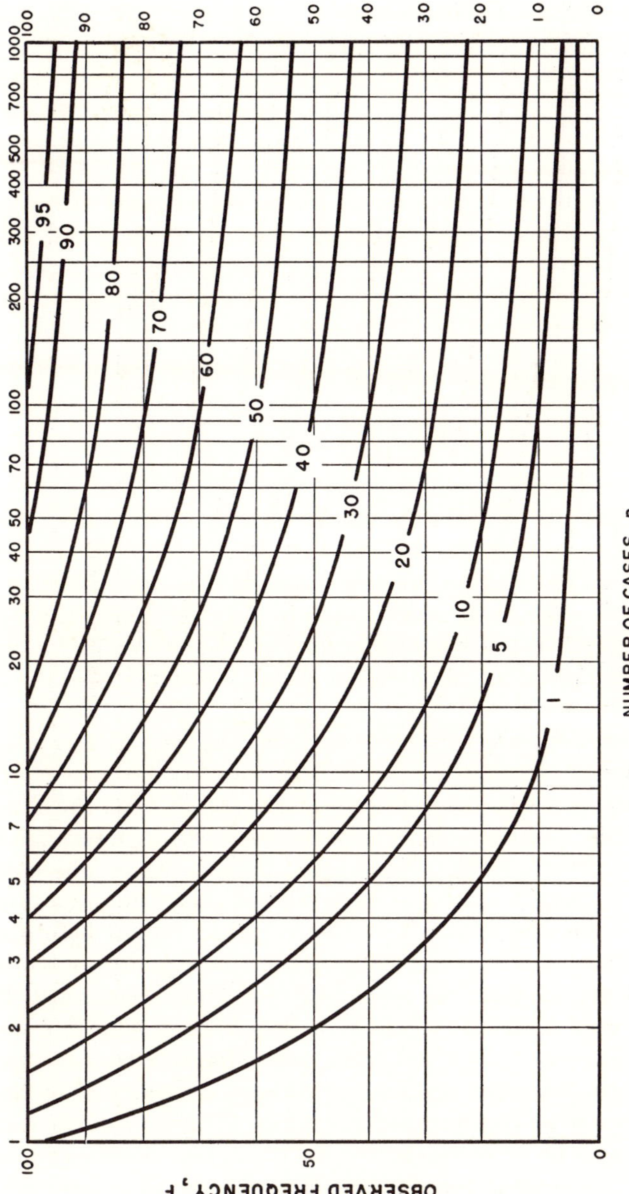

FIG. 3. Curves of the lower 95 % confidence limit of the true relative frequency of an event that occurs with observed relative frequency *F* in *n* independent cases.

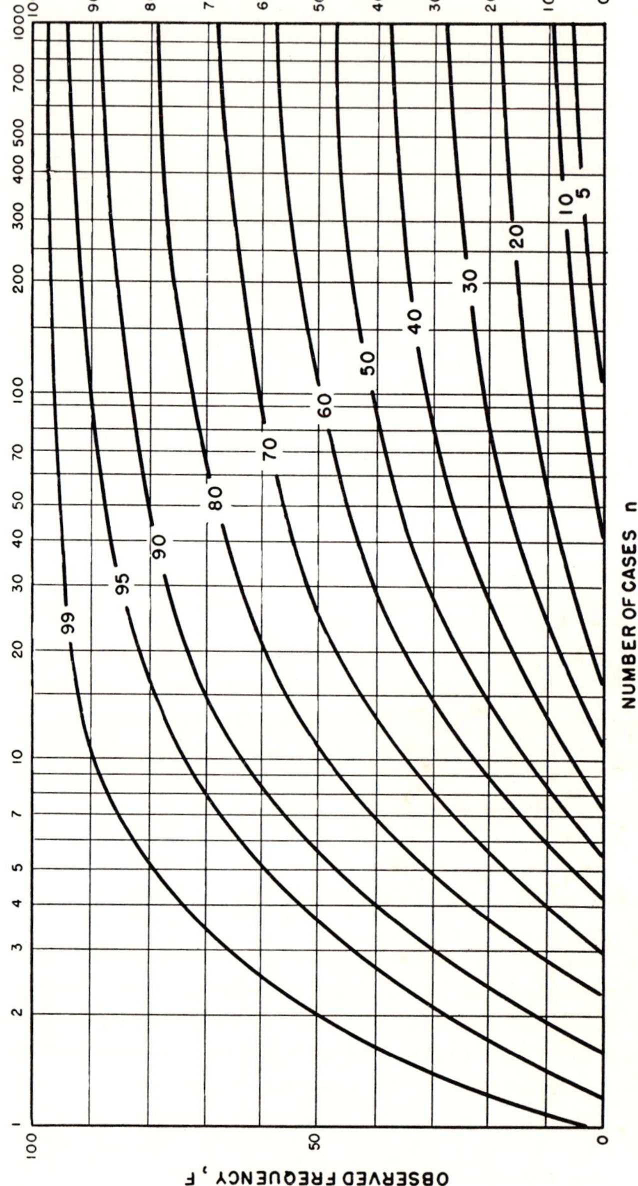

Fig. 4. Curves of the upper 95 % confidence limit of the true relative frequency of an event that occurs with observed relative frequency F in n independent cases.

TABLE I. Part of a contingency table between predictors A, B, C and a predictand X. Each n represents the number of observed cases in its cell.

Predictors			Predictand					Marginal Total
			X_1	X_2	X_3	X_4	X_5	
A_1	B_1	C_1	n_1	n_2	n_3	n_4	n_5	T_1
		C_2	n_6	n_7	n_8	..	..	T_2
		C_3	..	..	..	..	..	.
		C_4	.					.
	B_2	C_1	.					.
		C_2	.					
		C_3						
		C_4						
A_2	B_1	C_1						
		C_2						
		.						
		.						
		.						
								N

which the combination of events was observed (the n's in Table I), or might give the relative frequency with which each predictand class follows each combination of predictor classes (ratio (8) above), or might give the *contingency ratio* for that cell [47], which, symbolically, is

$$\frac{(A_qX_k)}{(A_q)} \Big/ \frac{(X_k)}{N} \tag{9}$$

between predictor class A_q and predictand class X_k for an over-all sample of N days.

There have been several studies to measure the association between predictors and a predictand [46, 51, 52]. Woodbury and Holloway [53] refer to I in the following expression as the "information ratio," a number ranging between zero and unity; the greater the information provided by the predictors, the greater is the information ratio.

$$I = \sum_a \sum_k (U_aX_k) \log \frac{N(U_aX_k)}{(U_a)(X_k)} \Big/ \sum_k (X_k) \log \frac{N}{(X_k)} \tag{10}$$

where U_a represents $A_qB_rC_s \cdots$ An "index of efficiency," also ranging between zero and unity, which my colleagues and I have used for several years, is

$$I = \left[\sum_a \sum_k \frac{(U_a X_k)^2}{(U_a)N} - \sum_k \frac{(X_k)^2}{N^2}\right] \Big/ \left[1 - \sum_k \frac{(X_k)^2}{N^2}\right] \tag{11}$$

There is a technique that is a combination of Devices V and III which has not been given wide distribution but is of academic, if not practical, interest [54]. Contingency tables are drawn between a predictand and each of a set of predictors, from which the probability of each class of the predictand following each class of each predictor can be estimated. Then, by a statistical device resembling the method of "least squares," a weighted probability of each predictand class can be obtained. Symbolically,

$$p(X_k|A_qB_rC_s \cdots) = \sum_R w_i p(X_k|R_i) \tag{12}$$

where R_i represents, in succession, A_q, B_r, C_s, . . . ; $p(X_k|R_i)$ is the conditional probability that the predictand class X_k will follow the ith class of the predictor R; and $p(X_k|A_qB_rC_s \cdots)$ is the probability that X_k will follow the combination of the observed classes of A, B, C, . . . The w's are weighting factors subject to the condition that their sum is unity. The greater the association between the predictor A and the predictand X, the greater should be the coefficient w_q of $p(X_k|A_q)$. But the size of each w is influenced also by the amount of association between the predictors themselves; when two predictors give essentially the same information, their probabilities are weighted so as to reduce their influence to that of one predictor.

The above technique was designed to avoid very small samples when estimating $p(X_k|A_qB_rC_s \cdots)$. Wahl [47], in an effort to avoid the same difficulty, uses the product of all of the contingency ratios between A_q, B_r, C_s, . . . and X_k. Symbolically, a quantity P is defined by

$$P(X_k|A_qB_rC_s \cdots) = \frac{(A_qX_k)N}{(A_q)(X_k)} \cdot \frac{(B_rX_k)N}{(B_r)(X_k)} \cdot \frac{(C_sX_k)N}{(C_s)(X_k)} \cdots \tag{13}$$

If $P(X_k|A_qB_rC_s \cdots) > P(X_m|A_qB_rC_s \cdots)$ then X_k is accepted as a more likely event than X_m following the antecedents A_q, B_r, C_s, ...

2.1.6. Device VI. The Analogue. Device V, continued until the subsample consists of only one case, reduces to an analogue technique. Frequently the predictors are successive classes of an attribute [54], such as the successive positions of a low-pressure center. If a past sequence of weather maps can be found which is analogous to the present sequence, then the previous sequence constitutes an analogue of the present sequence.

Recently an analogue technique was developed by Grytøyr [55] in which he represents on a continuous time scale the movements of fronts, air masses, low-pressure centers, and high-pressure centers. An analogue is selected which resembles the present situation most closely in the sequence of developments. Perhaps this will not be generally considered an objective technique owing to the strong appeal to subjective intuition and reasoning.

On analogues, Douglas [17] writes, "As far back as the 1890's the method has been tried and found wanting. The experience on which the individual forecaster builds up his technique is based on the normal modes of behaviour in a large number of somewhat similar situations."

2.1.7. Device VII. List of Rules; Stratification of Cases; Process of Elimination [56, 57, 58, 59]. This technique is, naturally, an old one. The forecast rules are verbal, and might read like an essay summarizing the investigator's experiences. But, basically, the technique resembles Device V in directly associating predictands with predictors, although the predictors might be obscured by the verbiage. Instead of graphs or tables, the forecaster might be instructed by the author of the method to use a worksheet [60, 61].

Symbolically, the technique might work as follows:

(a) The forecaster determines whether the condition A prevails or not-A. If not-A, then he forecasts not-X.

(b) If A prevails, then he determines whether B prevails or not-B. If not-B, then he forecasts not-X.

(c) If B prevails, then he determines whether C prevails or not-C, and so on.

By a process of elimination this method could isolate days when there is a strong possibility of X (a tornado, say). This technique may, or may not, be accompanied by supporting data. It is usually presented as a method of forecasting events categorically.

Grouped with this technique are several recent papers that attempt to outline, through physical reasoning, the regions of uplift or vertical motion [62, 63]. Wherever the motion is upward precipitation is expected. In other words, a single parameter, representing vertical motion, although difficult to derive, is used to answer "yes or no" to the question of precipitation.

2.1.8. Discussion. The above classification of techniques has placed emphasis on the statistical and climatic approach to forecasting rather than upon the physical or scientific numerical approach. The classification could serve hydrometeorological forecasting as well [64, 65]. Landsberg and Jacobs [66], in fact, have defined objective forecasting as "the use of the synoptic-climatic method for weather forecasting." Each of the seven devices described above is not generally used exclusively in any one study; the investigators could not be expected to think in my terms. For example, the scatter diagrams of Device IV might be grouped into two or three sets, each set applicable and answering to a preliminary description of antecedent weather (Device VII) [67, 68, 69]. The distinction may be thin between an empirical equation (Device II) and a regression equation (Device III) [70, 71].

Classification of objective techniques could have been accomplished in several other respects. Some techniques yield a probability figure for each possible development. Devices I, II, VI, and VII are used primarily to provide categorical answers to operational problems; Devices III and IV can be used either way; Device V, however, is more useful in providing probability statements on future developments [72].

There has been no attempt here to categorize the investigations themselves, to stratify the research itself according to the subjectivity or objectivity of the investigations. (A purely objective method of forecasting could have been assembled in a completely subjective manner.) No mention has been made of the difficulties of collecting and processing data, or of the computing difficulties. Systems differ in the volume of work required to prepare them; some are facilitated by the use of business machines.

2.2. Predictors

Far more than the device, the choice of predictors will determine the success or failure of an objective forecast method. Forecasting can be no better than the meteorological science that enters into it. Failure to recognize this fact might well account for the failure of many objective studies that have been based on hit-and-miss methods of selecting parameters for correlation or association. It is gratifying, therefore, to note in most recently published studies that the physical reasoning is given for the selection of the predictors.

The determination of a predictor itself often involves a technique. For example, when the temperature upstream from a station is used as a predictor of the maximum temperature, it cannot be found until the author of the objective system describes his technique for finding the upstream temperature. The gathering of the predictors might be simple or difficult, depending upon their complexity. The following is a list of predictor types, roughly in order of complexity:

(a) A *direct observation*, such as the present weather, present dew point, wind direction, and so on.

(b) A *tendency* or *gradient*, such as the change of sea-level pressure during a three-hour period or the pressure gradient over a given point.

(c) An *extrapolation*, which differs from a tendency by its being projected into the future.

(d) A *weather-map analysis* or *type*. (The typing of a prognosticated chart could serve as a predictor.)

(e) A *derived* parameter such as one derived from the analysis of an upper-air sounding of the pressure, temperature, and humidity to give the potential vertical velocity inside the air mass.

(f) A *computed* parameter, computed by formula from basic measurements.

2.2.1. Direct Surface Observations. Of some 30 papers written in 1952 and 1953, less than half have used direct surface observations, perhaps indicating that the investigators prefer more sophisticated tools. The moisture content of the surface layers [26, 31, 70, 73, 74], whether it is measured as dew point, temperature minus dew point, wet-bulb temperature, or relative humidity, has been the most frequently used attribute of the local surface weather, followed by temperature, wind direction, and cloud cover. The existing condition of the predictand, such as rain, is not used as often as it might be, perhaps because most objective studies are aimed at forecasts longer than 24 hours in advance. Sea-level pressure and pressure tendency are noticeable by their absence in objective studies, even though a "falling barometer" has been the most celebrated simple predictor of approaching bad weather.

2.2.2. Upper-Air Observations. The soundings of upper layers of the atmosphere have enjoyed frequent application, in spite of their inaccuracies and their failure to be representative of rapidly changing air masses. The methods by which a sounding is analyzed are as numerous as the analyzers, but this fact is largely due to the different purposes for which a sounding is used [e.g., 75]. In the last two years Shuman and Carstensen [36], Beebe [48], Means [69], and Schmidt [59] have used the lapse rate of temperature (e.g., temperature at 850 mbars less temperature at 700 mbars) to provide the clue to the stability of the air mass for tornado, thunderstorm, or precipitation forecasting. However, Fawbush and Miller [58], Showalter [28] and Gringorten and Press [71] have preferred to view upper-air instability as due to the density contrasts that might develop between moist rising air and the entrained drier air. Gilbert [76] and Williams [27] have considered upper-air temperature and humidity useful in forecasting surface minimum or maximum temperature. In fog predictions [77], the aerological soundings seem to have been set aside, at least in the literature. But a relatively new and successful use has been made of the sounding to predict condensation trails [19]. Its value in forecasting aircraft icing is almost self-evident [30].

2.2.3. Surface Weather Charts. The surface weather charts have been in favor and disfavor among objective forecasters since their first use in 1885 [78], largely because their use demands a compromise with subjectivity. Surface weather maps give the trajectory of air parcels for such predictands as maximum temperature [79] or give the position of low centers [74, 80] or give the surface pressure gradients. In some studies the difficult task of typing the weather maps has been tackled [73, 81, 82], whereas in other studies the weather maps are used negatively to eliminate cases in which the changes in air mass would upset the forecast, say, of minimum temperature [83].

2.2.4. Upper-Air Charts. The upper-air charts, at 850 mbars, 700 mbars, 500 mbars, and 300 mbars (roughly at 5,000 ft, 10,000 ft, 18,000 ft, and 30,000 ft) are becoming more popular each year with objective investigators. A survey, in 1949, of opinions of forecasters themselves [84] placed the 700-mbar upper-air charts at the top of a list of sources of forecasting information. These charts, usually more conservative and simpler than the surface charts, lend themselves to typing and to the smoothing of lines of equal pressure and temperature. Further, the physical, dynamic, and synoptic literature of the past 10 years have added much to the meteorologist's understanding of the atmospheric processes interrelating upper-air and surface movements. The flow of air in the upper levels, with its associated temperature advection, has been used in the investigations of maximum and minimum surface temperatures [57], thunderstorm and related phenomena [21, 36, 58, 74], and turbulence [85], as well as in the study of ordinary rainfall [41, 60]. But the rules and prescribed methods of analysis of the charts vary considerably, each author striving to approach that fabulous goal: strict objectivity. Some 25 rules have been examined critically by Harrison [56] and found generally useful. The movement of moisture aloft has, to date, not received the attention it would seem to merit [86].

2.2.5. Number of Predictors per Study. In 1946, Brier [67] used some 14 predictors in his classical study using the graphical technique. In 1950, Thompson [41] used six parameters. In 1951, Schmidt [60] made one fortunate choice for a predictor that, by itself, served to increase measurably the accuracy of forecasts of precipitation in the Washington, D. C., area. There has been a trend in the methods to depend upon a few well-founded predictors rather than upon many predictors, a trend which it can be hoped will continue.

2.3. Predictands

This report deals only with predictands that are expressed in operational terms, such as amount of precipitation, cloud cover, minimum temperature, or fog. I omit studies that lead only to the prognostication of surface and upper-air pressure patterns.

The composer of an objective forecast system might experience considerable frustration with the predictands. For example, he might begin with the intention of studying the forecast of precipitation: incidence, amounts, and kinds. Soon it will appear that distinguishing between kinds of precipitation is a study in itself. The investigator will have to choose between forecasting the incidence of precipitation for a given instant and forecasting amounts in an interval of time. Also he must decide at what station, or in what region, rain must occur for verification. There is the problem of what constitutes a rainfall. Some investigators will classify a *trace* as no rain at all; others will consider it as a verification of rain. In the case of the knotty problem of fog at an airport there are the difficult questions of when the fog begins and ends, how intense it is, whether a fog bank at the end of a runway constitutes a legitimate observation of fog at the station, and so on.

The titles of many objective studies bear mute evidence on the difficulty of increasing the predictand information: "Tornado forecasting for Kansas and Nebraska" [36], "Precipitation at Washington, D. C., during month of May" [59], "Rain or Snow at Denver, Colorado, September–November" [37], and so on. Some predictands are better suited to objective studies than others. Thus, many early reports centered in the forecasting of the frost danger to crops [29]. In the past two years tornados have received much attention by objective forecasters [28, 36, 58].

An investigator might describe a predictand in one of the following ways:

(a) A *continuous variable* such as temperature, rainfall amount, visibility (in miles).

(b) An *attribute* dividing the days into mutually exclusive classes such as "clear, partly cloudy, and overcast."

(c) A *dichotomy*, or a simple choice such as between "rain" and "no rain." This includes an isolated event or phenomenon of extreme nature such as a tornado, thunderstorm, or frost, to which the objective system must provide the answer "yes" or "no."

Objective studies of the last several years, grouped by the predictands, are as follows:

Continuous variables
- minimum temperature (frost included) [22, 23, 26, 31, 57, 70, 76, 83, 87]
- maximum temperature [88, 79, 27, 49]
- precipitation (amount) [67, 24, 89, 90]
- hailstone size [21]
- relative humidity [91, 92]
- cloud ceiling [93]
- cloud top [94]
- upper-air turbulence [71, 85]

Attributes in classes
- ceiling [95, 73]
- visibility [73]

Dichotomies
- precipitation (occurrence) [96, 40, 62, 59, 60, 63, 25, 80, 82, 73, 41]
- thunderstorms [48, 69, 28, 50]
- snow [97, 98, 99, 39]
- rain-or-snow [37, 100]
- tornado [74, 58, 36]
- fog or stratus [35, 68, 101, 102]
- ice fog [20]
- frost (see under minimum temperature) [103]
- northerly winds (for fire control) [104, 105]
- gusts [73]
- condensation trails [19, 106]
- icing of aircraft [30]
- cirrus clouds [107].

2.4. Newer Techniques

The problem of increasing the predictand information was admirably attacked by the Short Range Forecast Development Section of the U. S. Weather Bureau [108]. By the scatter-diagram device some five predictors are used to classify a given weather situation (designated, say, as IA). Then by means of a set of nomograms (e.g., Fig. 5) the simultaneous values of two continuous parameters (Y_c, Y_1) are related to an observable predictor (A_0). The predictor A_0 might be the initial height of the clouds, Y_c the height of the clouds six hours later, and Y_1 the probability that the clouds will be at that height. Or Y_1 might be the time at which the clouds are predicted to lower to the new height Y_c. Or Y_1 might be the duration of a cloud cover at the height Y_c or lower. The system as presented is strictly objective in its application.

Recently, at the U. S. Air Force Cambridge Research Center, we have felt that a statistical aid will have greater appeal to a practicing forecaster if he can learn what historical cases enter into the statistics. He might prefer to study a sample of cases to help him visualize the subsequent developments. We therefore devised a *register* [73] to serve as an aid in forecasting (Fig. 6). Each line in the register is devoted to the weather of one day of a long record of days, say 10 years. In Fig. 6, representing part of one page in a register, the group of six figures near the middle of each line gives the date for that line. To the left of the date are 25 columns of predictor data, in coded form. A *reference guide* is provided the forecaster to interpret the codes. To the right of the date are 20 columns that describe the three-hourly weather, in coded form, beginning with the weather at the initial period (0830E in the illustration) and covering the developments up to 30 hours later. Examples of the predictand codes, X and Y, are shown in Tables II and III, and represent

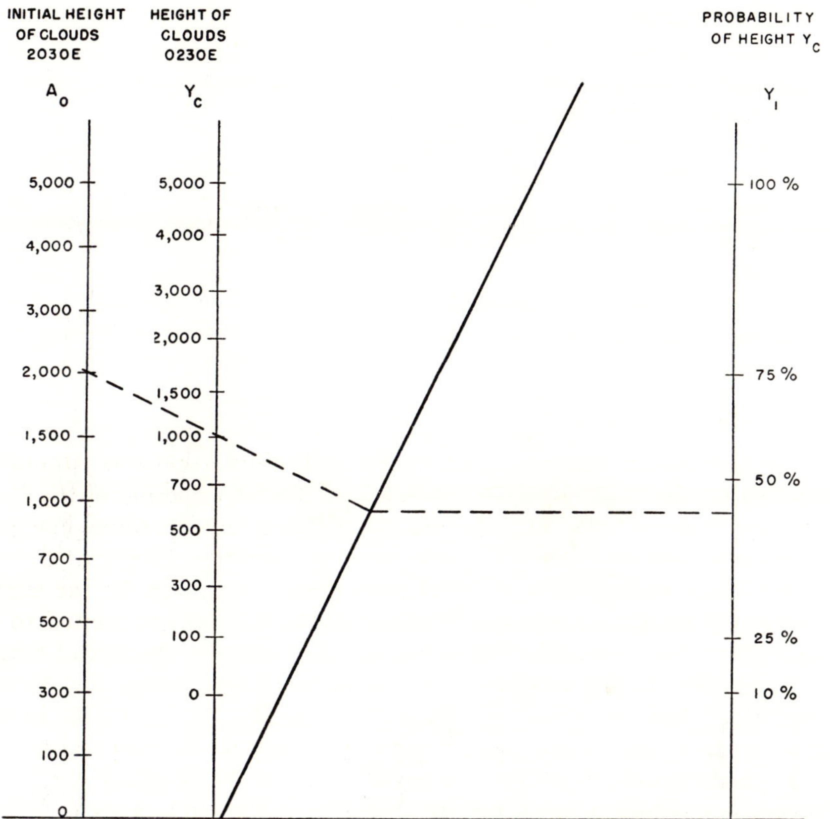

FIG. 5. A nomogram for antecedent conditions (I.A.) to relate simultaneous values of two parameters (e.g., Y_c = 1,000 ft, Y_1 = 45 %) to a predictor (A_0 = 2,000 ft).

an effort to cover several events of operational importance: cloud heights, visibility, kind of precipitation, and gusts.

For sampling the past weather, the register is arranged in order of a key predictor, such as the weather-map type. The forecaster selects lines in which the predictors resemble the predictors of the present day. If the forecaster isolates a sample of, say, 15 days, he can determine with a fair degree of accuracy the probability of each subsequence at three-hour intervals. But, instead, he might scrutinize the various develop-

PREDICTORS																									DATE			PREDICTANDS																			
WEATHER MAP ANALYSIS		LOW CENTER DISTANCE	HIGH CENTER DISTANCE	FRONT OR TROUGH DISTANCE	6-HR MOVEMENT OF LOW	DIRECTION OF MOVEMENT OF LOW	POSITION OF LOW W.R.T. N.Y.	TEMPERATURE TENDENCY	NUMBER CLOSED ISOBARS	6-HR MOVEMENT OF FRONT	DIRECTION OF FRONT W.R.T. N.Y.	TEMPERATURE - DEW POINT	LATEST OBSERVATION BELOW MINIMA	DAMPNESS OF THE GROUND	TEMPERATURE	PREVIOUS POSITION OF AIR PARCEL	WIND DIRECTION 1000 FT.	LOWEST CLOUDS	SURFACE WIND DIRECTION	WIND VELOCITY 3000 FT.	WATER TEMPERATURE - DEW POINT	2-HR SKY COVER CHANGE	WEATHER AT 0830 E	GUSTS/ OBSTRUCTIONS 0830 E				1130 E		1430 E		1730 E		2030 E		2330 E		0230 E		0530 E		0830 E		1130 E		1430 E	
1	2	3	4	5	6	7	8	9	10	11	12	13	14	15	16	17	18	19	20	21	22	23	24	25	YR	MO	DAY	X	Y	X	Y	X	Y	X	Y	X	Y	X	Y	X	Y	X	Y	X	Y	X	Y
3	4	5		3	2	4	7	1	2		9	1	2	7	3			1	0		5	0	6	6	45	12	31	6	6	6	6	6	6	2	2	2	0	4	0	2	0	4	0	4	3	4	3
3	4	8	7	6			6	4	2	3	9	0		9	2	6		2	1		3	0	7	7	47	12	26	7	8	7	8	7	8	4	8	7	8	7	8	3	3	3	0	2	8	0	0
3	5	9		1	1	5	7	2	0	3	9			2		1	1	7	1			6	2	0	47	1	27	0	0	0	0	0	0	0	2	0	2	0	2	0	2	2	2	2	2	0	2
3	5	8		0	2	3	7	7		1		2	1	1	4			8	8			7	2	4	45	2	16	2	2	2	2	2	0	0	2	2	0	3	0	3	1	4	0	4	0	4	2
3	5	9		0	2	3	7	0	0	3	1	2		6	2			3	9		4	5	3	3	46	2	20	2	1	2	3	0	3	0	3	0	3	0	3	0	3	0	3	0	3	0	3
3	5	9			2	2	7	1				1		4	4	3		2	6		8	9	6	6	43	2	11	6	6	6	0	7	8	0	3	0	3	0	0	0	0	0	0	2	0	0	2
3	5	9		1	1	2	7	5	7	0	1	0	0	9	4	4		1	6		8	0	5	5	43	3	17	5	5	6	6	6	6	3	4	1	4	1	4	0	4	1	1	1	1	2	3
3	5	9	7	1	1	2	6	3	9	3	1	0		7	3	7		2	2		5	0	6	6	44	3	30	6	6	6	6	4	0	4	5	0	0	0	0	4	5	4	2	2	0	2	3
3	5	9		1	2	2	7	8			9	5		0	6	2		7	7		6	6	2	2	43	5	7	0	2	2	3	0	3	1	0	0	0	2	0	2	2	0	0	0	2	0	2
3	5	6		0	2	3	7	6	2	2	2	1		8	5			2	5		4	0	5	5	46	5	19	2	1	5	1	0	1	0	2	0	2	0	2	0	2	1	2	2	2	3	0

FIG. 6. The upper part of one page of a register. Each line consists of antecedent information, date for the line, and subsequent developments in the weather of the station.

TABLE II. A weather classification of sky cover and precipitation.

Symbol	Item
X_0	Clear or scattered clouds
X_1	Thin high broken or thin high overcast
X_2	Broken
X_3	Overcast with ceiling $\geq$ 5,000 ft
X_4	Overcast with ceiling 1,000 to 4,900 ft
X_5	Overcast with ceiling (including fog) $<$ 1,000 ft
X_6	Rain or drizzle (including showers)
X_7	Snow (including snow showers, with or without rain)
X_8	Freezing rain or sleet* (with or without rain or snow)
X_9	Thunderstorm and/or hail*

* Observed at time of verification or within three hours following the time of verification.

TABLE III. A classification of gusts, winds $\geq$ 25 mph, and/or obstructions to vision.

Symbol	Item
Y_0	None (visibility $>$ 6 miles, no gusts, winds $<$ 25 mph)
Y_1	Haze
Y_2	Smoke (with or without haze)
Y_3	Gusts or surface winds $\geq$ 25 mph with visibility $<$ 3 miles observed at time of observation and during the 3 hours following verification (with or without smoke or haze)
Y_4	Ground fog (with or without smoke or haze)
Y_5	Fog (with or without smoke or haze)
Y_6	Fog with precipitation (with or without smoke or haze)
Y_7	Gusts or surface winds $\geq$ 25 mph at time of verification or within three hours following verification *and* fog at time of verification
Y_8	Gusts with any obstruction to vision other than fog, or an obstruction whipped up by the wind; visibility $<$ 3 miles at time of verification or within 3 hours following
Y_9	Precipitation reducing visibility to $<$ 3 miles

ments in his sample to alert himself to the possibilities during his tour of duty, and make an educated effort to predict what will really happen.

3. The Success and Future of Objective Forecasting

3.1. *Determining Success*

The success of objective methods, to date, is evidenced by the prevalent use of objective forecasts of frost and the increasing number of papers on many other predictands. Many objective tools remain unpublished or in manuscript form [88, 92, 99, 109]. Their use is restricted to local offices and encouraged by forecasting or research agencies, including the U. S. Weather Bureau, U. S. Air Weather Service [110], the Meteorological Division, Department of Transport, Canada, and the Meteorological Research Committee, British Air Ministry.

Forecast improvement, however, has not been so overwhelming as to be obvious without a systematic program for the comparison of techniques. Yet there has not been an acceptable verification program of long standing by which to judge the performance of forecasters. Claims and counterclaims should stand or fall on verification, a subject that has not been neglected in the last three years [111–115].

Of all the papers mentioned in the references below, only a few present the comparisons of results of forecasts by the objective methods with results by conventional methods. The largest improvement is indicated by Schmidt [60]. In his study of the occurrence of winter precipitation at Washington, D. C., he shows that, whereas the official forecasts usually yielded 60% to 70% of the forecasts correct, the systematic forecasts were correct on 85% to 90% of the same sample of days.

Schmidt's indicated improvement is large enough to be independent of the choice of method of verification. But in other studies the improvement can be questioned. In Schmidt's other study of the occurrence of precipitation in the month of May [59], the number of correct forecasts in a test sample of 53 days was increased by the objective system from 80% to 90%. His sample should have been twice as large to make this apparent improvement significant [116].

The investigators of the U. S. Weather Bureau [108] have indicated that the systematic forecasting of occurrence of cloud heights at or below 5,000 ft improves the probability of correct forecasts from 65% to 75%. In the same study, systematic forecasting of the occurrence of visibility of 5 miles or less increases the number of correct forecasts 10% to 15%, to give approximately 75% of these forecasts correct. Thompson [41], in studying rainfall amounts in Los Angeles, California, finds that the degree of accuracy was close to 80% in the selection of one of five categories of amounts, a slight, but encouraging, edge in favor of the objective system.

Some authors, with recognizable justification, have expressed difficulty in comparing the results of their systematic forecasts with the efficiency of conventional forecasts. Whereas an objective system must yield a forecast, say, of thunderstorm for a given time interval, in a well-defined area and for a specified number of hours in advance, a conventional forecast might loosely state "scattered showers possible" in and around a city or State. Nevertheless, Beebe [48] is able to state that the degree of accuracy by his systematic methods of 12- to 24-hour forecasts of thunderstorms, at one or more of five specified stations in Tennessee, Alabama, and Georgia, is nearly the same as by conventional forecasts. His degree of accuracy for categorical statements is 80%. But Beebe also gives the numerical probability of thunderstorms with each forecast. Means [69], in a similar study, obtained similar results. Gilbert's objective system [76] for the forecast of minimum temperature at Denver, Colorado, performed almost as well as the subjective system of the official forecasts, yielding approximately 70% of the forecasts correct within 5° F. But Gilbert raises an interesting point by noting an improvement in the official forecasts during the period within which the objective system was tested. The forecasters became familiar with the objective system and were "able to improve on the objective indication of the forecast."

Many investigators have developed their objective systems by using the data of one period, and have tested the stability of the results in an independent period. The justification for publishing these devices as tools for future forecasting rests primarily upon the quality of reiteration of the results. The forecaster who uses such a device will know what degree of accuracy to expect from its use, and will set the device aside if he can obtain better results by another method. The same argument holds for the papers wherein the authors develop the forecasting tools from basic physical theory and test the device on historical samples [21, 24, 25, 27, 62, 63]. The chief merit of some papers, however, is in calling the attention of forecasters to the schemes they present for the forecasting of perplexing events [19, 28, 30, 57, 93], for example, icing aloft.

3.2. The Degree of Accuracy

A figure that is frequently given for the degree of accuracy of forecasts is 80%. According to Douglas [17], even before 1914 the weather services were quoting a figure like 87%, stating that this percentage of forecasts gave "fairly accurate indication of weather." The degree of accuracy, however, can be adjusted up or down simply by the relaxing or restricting of the requirements for verification. For this reason the following paragraphs give, as much as possible, the requirements that are met by the forecast method, together with the degree of accuracy.

In the forecasting of minimum temperatures, mostly for frost danger, Jacobs [23] indicates that the forecasts, in the absence of strong winds, are correct within 1 to 3° C on clear nights and 1 to 4° C on cloudy nights. Lessmann [26] and Kern [31] give figures to show that, in 70% to 80% of the cases of "radiation nights," the error of forecast of minimum temperature does not exceed 1.5° C. Williams [27], using insolation theory, indicates his degree of accuracy to be approximately 1° F for 50% of his forecasts and 2° F for nearly 70% of the forecasts. Saunders [101] shows that the temperature of fog formation can be predicted within 1° F in 90% of the cases. Johnson [57] performs his service as an author in *alerting* forecasters to conditions when frost is "likely."

On precipitation amounts, Brier [67] devised a system to predict winter precipitation in five categories from no precipitation to heavy precipitation in the TVA Basin during a 24-hour period immediately following the time of forecast. He correctly predicted 44% of the categories, missed another 44% by one category, making gross errors in only 12% of 89 trial forecasts. Richards [89], using visibility in falling snow at some Canadian stations, indicates 85% of his predictions correct for the hourly rate of accumulation, predicting light snowfalls within 0.2 inches and heavy snowfalls within 0.5 inches. Jorgensen [117], in relating the synoptic charts to precipitation amounts, finds that his system yields 85% of his "forecasts" correct.

As in Jorgensen's paper, precipitation has been linked by others to synoptic charts [24, 25, 40, 62, 63], in preparation for the day when prognostic charts will be greatly improved. Teweles and Forst [40] find that rain-or-no-rain forecasts, 36 to 48 hours in advance, can be correct roughly 80% of the time, the same as it is for the official forecasters. Kuhn [25] and Riehl [62] have delineated expected areas of precipitation, but it is difficult to comprehend the degree of success. To determine whether precipitation will be rain or snow, Laird and Dickey [37] have published results showing correct choices in 95% of their cases, 14 to 26 hours in advance. Murray [100], using only the thickness of the 1000–700-mbar layer aloft, could correctly distinguish 92% of the cases between rain and snow, not classifying the borderline cases of sleet either way.

The forecasting of cloud heights is difficult, owing to the variability of these heights in fog and precipitation. Goldman [95] has predicted the height of clouds in precipitation within 200 feet, correct approximately 60% of the time. In my study of fog and stratus at Randolph Field, Texas, categorical forecasts of the occurrence of ceilings below 2000 ft during the night, 16 hours in advance, are expected to be correct in 70% to 86% of all cases. The probability of each occurrence could be given within a 10% range.

Fawbush and Miller [21] have shown how to predict hailstone sizes within 1/16 inch for the diameter of the smallest hailstones and within 1½ inches on hailstones as much as 4 inches in diameter. In the verification of some 219 forecasts they were able to obtain 77% of the forecasts correct. The errors were usually on the large side.

In their study on airport acceptance rates [108], the investigators of the U. S. Weather Bureau indicated that their method of predicting wind speeds, two, four, and six hours in advance, were correct in 95% of the cases, within 12 miles per hour for light winds and 25 miles per hour for strong winds. They used a physical approach to forecast speed, and suggested that an empirical investigation might give a higher degree of accuracy. Their predictions of wind direction were poor. Upper-level gust velocities [71] can be predicted within 10 ft/sec in 95% of the cases and within 5 ft/sec in 67% of the cases. Such quantitative figures are not likely to be given except in an objective technique.

On aircraft icing, both the Air Weather Service [30] and the U. S. Weather Bureau [108] give information that does little more than alert an operational office to the conditions favorable to this danger. Appleman's methods [19, 20] for predicting the formation of condensation trails aloft and ice fogs at the surface in high latitudes are derived from basic theory and are in general use [106].

The destructiveness of tornadoes is all out of proportion to their frequency. Their scarcity makes it difficult to test a forecasting method statistically. Shuman and Carstensen [36] indicate that, by 9:00 a.m. the forecasting of one or more tornadoes in an area in the Middle West, roughly the size of Kansas, has been correct for 60% of such forecasts. No-tornado forecasts have been correct in 99% of such forecasts. Forecasting for the State of Georgia, Armstrong [74] indicates success in 36% of the tornado forecasts. In presenting an index of thunderstorm and tornado possibilities, Showalter [28] prefers to state that a high index value is "very likely to be associated with . . . " thunderstorm and tornado activity.

The above information on the accuracy of objective forecasts must remain sketchy because of inherent difficulties in appraisal. The number of correct forecasts *per se* can be meaningless [118]. In any one study the same number of correct forecasts conceivably might have been made by methods of chance or simple rules of thumb. The results of an investigation can be judged from one of several points of view, of which the viewpoint of accuracy might be the least important. Forecasting might be judged for *skill* or for the *utility* of the forecasts with respect to specific operations [111, 112]. But the subject is beyond the intended purpose of this review.

3.3. The Future of Objective Methods

It is clear that an objective system of forecasting can be no better than the meteorological skill that enters into its construction. Research into the physics or dynamics of the atmosphere must continue if there is to be, some day, a greater degree of accuracy in weather forecasting. Yet no one can say with certainty that an improved model of the atmosphere and atmospheric processes will result in forecast improvement. Instead, an increasing knowledge might lead to a better understanding of the inherent unpredictability of the weather. According to Sutcliffe [13], "A form of statistical physics applicable to meteorological prediction seems to be required."

There is a trend among investigators in objectivity to relate operational weather to current synoptic weather patterns in preparation for the day, if it should come, when prognostic charts will be highly improved [40, 117]. Thompson, Collins [24], and Kuhn [25] use a formula derived by J. R. Fulks in 1935, which relates rate of precipitation to vertical motion. They choose a method of computing vertical velocity, correct for moisture in the air, and estimate rainfall amounts and rainfall patterns. Sawyer published a similar study in 1952 [119]. Riehl, Norquest, and Sugg [62] derived a formula for the sign of the vertical motion at an upper level (300 mbars) and show how well the fields of positive vertical motion compare with concurrent precipitation areas. In a later paper, however, Teweles [63] tests this predictor, along with two other plausible indicators of vertical motion, and concludes that their relation with precipitation "is not close enough."

And what of the efforts to improve the prognosis of surface and upper-air weather patterns, together with the related vertical motions? There are many recent papers in this field that exceed the intended scope of this report. The names alone, that are associated with these efforts, comprise a very impressive list including J. Charney, D. Fultz, E. T. Eady, A. Eliassen, R. Fjørtoft, P. D. Thompson, J. L. Sawyer, and F. H. Bushby, all of whom use a dynamic approach; F. Defant, University of Innsbruck, who uses a kinematic approach; J. J. George, Eastern Air Lines, Inc., and E. Rampey, U. S. Weather Bureau, who use an empirical approach.

For a rapprochement between dynamic meteorologists and statistically minded meteorologists who work on objective methods of forecasting, the latter will have to learn how to associate all operationally defined predictands with the parameters or patterns that the dynamicists learn to predict. For example, the *tornado* might be associated with a set of parameters that define a *tornado condition*, the parameters being numerically predictable.

Other factors, also, might influence the types of objective studies. Oliver and Oliver [120] believe that local meteorologists could apply objective techniques to the charts that are prepared in central offices and that are transmitted to the local offices by a "facsimile" process over telephone wires. The introduction of radar makes more pertinent the forecasts that *alert* an office to possible developments when the radar screen would be watched more closely. But the biggest factor will be the demands of the operational users. There have been two articles on air safety [121, 122] that enumerate the following predictands and explain their importance to aviation: frost, snow, icing, slush, rain, hail, blowing snow, fog, temperature inversions, lightning, gusts and turbulence, and slant visibility. Most of these predictands have been treated much too infrequently in objective studies (see section 2.3 above).

4. Concluding Remarks

There may be aspects of objective forecasting that have not been adequately treated in this review, or papers that have not come to the attention of the reviewer. Where should a study such as Spar's [123] have been inserted, which is a suggested but untried technique for quantitative precipitation forecasting? Have the efforts in the U. S. Weather Bureau as revealed in the Monthly Weather Review throughout the past 50 years, or in the Meteorological Department of Eastern Air Lines under the guidance of J. J. George been adequately covered? Even as this review is being edited another paper has appeared, which is an excellent example of curvilinear regression [124].

To begin with, this task was simplified by the "Bibliography on Objective Forecasting" issued by the American Meteorological Society in 1952 [3], which obviated the necessity for a comprehensive review of papers prior to that year. The purpose of this report, besides reviewing the latest papers, was to define the subject and its goals, to survey the attitudes toward, and compromises with, objectivity, to report on the degree of accuracy, and to speculate on its future. There will be, undoubtedly, many more objective studies, to keep pace with the ever-changing demands of the operational public who will often scoff at, but never dispense with, the forecaster.

Acknowledgments

I gratefully acknowledge the help of several people in calling my attention to one or more papers on objectivity: Mr. Malcolm Rigby, editor of Meteorological Abstracts and Bibliography, Mr. Conrad Mook, U. S. Weather Bureau, Mr. Robert Beebe, U. S. Weather Bureau, and my associate Mr. Iver Lund, Air Force Cambridge Research Center.

List of Symbols

$x_1, x_2, \ldots, x_n$	continuous variables, representing predictors
y	a continuous variable, representing a predictand
$f(x_1, x_2, \ldots, x_n)$	a function of $x_1, x_2, \ldots, x_n$
$a_1, a_2, \ldots, a_m$	empirically determined constants or parameters of the station
$b_0, c_1, c_2, \ldots, c_{12}, \ldots$	additional empirical constants
K_1, K_2	constant coefficients

p probability

$Y_1, Y_2, Y_3, Y_4, Z_1, Z_2$ intermediate quantities or predictors

W a predictand, either a continuous variable or an attribute divided into a finite number of classes

$F(x_1, x_2)$ symbol for a scatter diagram on which the variables x_1, x_2 are plotted to define a third variable (Y_1)

$F(x_1, x_2) \rightarrow Y_1$ represents the process by which the variables x_1, x_2 are related to a third parameter Y_1

$A, B, C, \ldots$ predictor attributes

$A_1, A_2, A_3, \ldots$ a finite number of mutually exclusive classes of the attribute A

$B_1, B_2, B_3, \ldots$ classes of B

$X_1, X_2, \ldots, X_k, \ldots$ mutually exclusive classes of the predictand X, representing a finite number of possible events

$A_qB_rC_s \cdots$ a sub-class of days on which the indicated combination of conditions occurred

$(A_qB_rC_s \cdots)$ the number of $A_qB_rC_s \cdots$'s in the period of years that have been studied

$(A_qB_rC_s \cdots X_k)$ the number of $A_qB_rC_s \cdots$'s that are characterized further by the subsequent event X_k

$(A_q), (X_k), (A_qX_k)$ the number of A_q's, X_k's, A_qX_k's, respectively

n number of days

N size of an over-all sample

U_a represents $A_qB_rC_s \cdots$

I a number ranging between zero and unity, termed either "information ratio" or "index of efficiency"

R_i represents, in succession, $A_q, B_r, C_s, \ldots$

$p(X_k|A_q)$ the probability that the event X_k will follow the condition A_q

$p(X_k|A_qB_rC_s \cdots)$ the probability that the event X_k will follow the indicated combination of conditions

$w_q, w_r, w_s, \ldots$ weighting factors, whose sum is unity

$\dfrac{(A_qX_k)}{(A_q)} \Big/ \dfrac{(X_k)}{N}$ contingency ratio between A_q and X_k in N cases

$P(X_k|A_qB_rC_s \cdots)$ product of the contingency ratios between each predictor class $A_q, B_r, C_s, \ldots$ on the one hand and the predictand class X_k on the other

References

1. Sutton, O. G. (1951). Mathematics and the future of meteorology. *Weather* **6,** 291–296.
2. Mook, C. P. (1949). A historical note on the development of systematic procedures for short range weather forecasting. U. S. Weather Bureau, Airport Station, Washington, D. C., 6 pp.
3. Kramer, Mollie P. (1952). Annotated bibliography on objective forecasting. *Meteorol. Abstr. and Bibliography* **3,** 471–517.
4. Bowie, E. H. (1906). The relation between storm movement and pressure distribution. *Monthly Weather Rev.* **34,** 61–64.
5. Lingelbach, E. (1952). Wird die Meteorologie eine exakte Wissenschaft? *Ber. deut. Wetterdienst US-Zone,* No. **38,** 7–9.

6. George, J. J. (1952). Objective forecasting studies. *Bull. Am. Meteorol. Soc.* **33,** 199–201.
7. Allen, R. A. (1949). Bibliography on objective weather forecasting. U. S. Weather Bureau Short Range Forecast Development Section, Washington, D. C., 8 pp.
8. American Meteorological Society (1951). Compendium of Meteorology. Boston, 1334 pp.
9. Willett, H. C. (1951). The forecast problem, *in* Compendium of Meteorology. American Meteorological Society, Boston, pp. 731–746.
10. Dunn, G. E. (1951). Short-range weather forecasting, *in* Compendium of Meteorology. American Meteorological Society, Boston, pp. 747–765.
11. Bundgaard, R. C. (1951). A procedure of short-range weather forecasting, *in* Compendium of Meteorology. American Meteorological Society, Boston, pp. 766–795.
12. Allen, R. A., and Vernon, E. M. (1951). Objective weather forecasting, *in* Compendium of Meteorology. American Meteorological Society, Boston, pp. 796–801.
13. Sutcliffe, R. C. (1952). Reviews of modern meteorology—6. Principles of synoptic weather forecasting. *Quart J. Roy. Meteorol. Soc.* **78,** 291–320.
14. Houghton, H. G. (1948). Some personal opinions of the present status of the science and practice of meteorology. *Bull. Am. Meteorol. Soc.* **29,** 101–105.
15. U. S. Air Weather Service (1952). Local Objective Forecast Methods Symposium, Jan. 23, 1952. Mimeographed notes. Washington, D. C., 16 pp.
16. Thompson, J. C. (1952). Applied research in weather forecasting. *Bull. Am. Meteorol. Soc.* **33,** 195–198.
17. Douglas, C. K. M. (1952). Reviews of modern meteorology—3. The evolution of 20th-Century forecasting in the British Isles. *Quart. J. Roy. Meteorol. Soc.* **78,** 1–21.
18. Probatov, A. N., and Shelegova, E. K. (1952). K. metodike predskazaniia srokov nachala neresta sel' di u zapadnogo poberezh'ia Iuzhnogo Sakhalina. *Meteorol. i. Gidrol. (Leningrad),* No. **5,** 51–53.
19. Appleman, H. (1953). The formation of exhaust condensation trails by jet aircraft. *Bull. Am. Meteorol. Soc.* **34,** 14–20.
20. Appleman, H. (1953). The cause and forecasting of ice fogs. *Bull. Am. Meteorol. Soc.* **34,** 397–400.
21. Fawbush, E. J., and Miller, R. C. (1953). A method for forecasting hailstone size at the earth's surface. *Bull. Am. Meteorol. Soc.* **34,** 235–244.
22. Brunt, D. (1939). Physical and Dynamic Meteorology. Cambridge University Press, England, 428 pp.
23. Jacobs, W. C. (1939). Application of Brunt's radiation equation to minimum temperature forecasts. *Monthly Weather Rev.* **67,** 439–443.
24. Thompson, J. C., and Collins, G. O. (1953). A generalized study of precipitation forecasting. Part 1. *Monthly Weather Rev.* **81,** 91–100.
25. Kuhn, P. M. (1953). A generalized study of precipitation forecasting. Part 2. *Monthly Weather Rev.* **81,** 222–232.
26. Lessmann, H. (1952). Frostwarnung und Frostschutz. *Ber. deut. Wetterdienst US-Zone,* No. **32,** 84.
27. Williams, T. L. (1953). The use of insolation in forecasting temperatures. *Bull. Am. Meteorol. Soc.* **34,** 245–249.
28. Showalter, A. K. (1953). A stability index for thunderstorm forecasting. *Bull. Am. Meteorol. Soc.* **34,** 250–252.
29. Zikeev, N. T. (1953). Selective annotated bibliography on frost and frost forecasting. *Meteorol. Abstr. and Bibliography* **4,** 337–397.

30. U. S. Air Weather Service (1953). Forecasting Aircraft Icing, Air Weather Service Manual 105–39. Dept. of the Air Force, Washington, D. C., 24 pp.
31. Kern, H. (1952). Zur Vorausbestimmung des Temperaturminimums in Strahlungsnächten. *Ber. deut. Wetterdienst US-Zone,* No. **42,** 107–109.
32. Schumann, T. E. W. (1944). An enquiry into the possibilities and limits of statistical weather forecasting. *Quart. J. Roy. Meteorol. Soc.* **70,** 181–195.
33. Wadsworth, G. P. (1953). Introductory work on a continuous prediction operator for upper-air pressure. Progress Report No. 9 (USAF Contract No. AF19-(122)-446). Cambridge, Mass., 5 pp.
34. Wadsworth, G. P. (1948). Short range and extended forecasting by statistical methods. U. S. Air Weather Service Technical Report, No. 105–38, 202 pp.
35. Gringorten, I. I. (1953). An objective system for estimating fog and stratus probability at Randolph Field, Texas. *Bull. Am. Meteorol. Soc.* **34,** 63–67.
36. Shuman, F. G., and Carstensen, L. P. (1952). A preliminary tornado forecasting system for Kansas and Nebraska. *Monthly Weather Rev.* **80,** 233–239.
37. Laird, H. B., and Dickey, W. W. (1953). Forecasting rain-or-snow at Denver, Colorado, September–November. *Bull. Am. Meteorol. Soc.* **34,** 287–292.
38. Nishida, M. (1953). Objective weather forecasting in Hiroshima. *J. Meteorol. Res. (Tokyo)* **4,** 627–630.
39. Fukuhara, K. (1953). On the snowfall forecasting. *J. Meteorol. Res. (Tokyo)* **5,** 85–88.
40. Teweles, S., Jr., and Forst, A. L. (1953). Forecasting winter precipitation 36 to 48 hours in advance at Des Moines, Iowa. An experiment using prognostic charts as a data source. *Monthly Weather Rev.* **81,** 357–367.
41. Thompson, J. C. (1950). A numerical method for forecasting rainfall in the Los Angeles area. *Monthly Weather Rev.* **78,** 113–124.
42. Bean, L. H. (1929). A simplified method of graphic linear correlation. *J. Am. Statistical Assoc.* **24,** 386–397.
43. Baur, Franz (1951). Extended-range weather forecasting, *in* Compendium of Meteorology. American Meteorological Society, Boston, pp. 814–833.
44. Diehl, H. (1952). Neue Mehrfachskorrelationstabellen zur Verwendung für Witterungsvorhersagen. *Meteorol. Runds.* **5,** 168–172.
45. Gringorten, I. I. (1949). A study in objective forecasting. *Bull. Am. Meteorol. Soc.* **30,** 10–15.
46. Gringorten, I. I. (1950). Forecasting by statistical inferences. *J. Meteorol.* **7,** 388–394.
47. Wahl, E. W., and collaborators (1952). The construction and application of contingency tables in weather forecasting. Surveys in Geophysics, No. 19. U.S.A.F., Cambridge, Mass., 37 pp.
48. Beebe, R. G. (1952). Thunderstorm forecasting in the Atlanta, Ga., area. *Monthly Weather Rev.* **80,** 63–69.
49. Kato, H. (1953). Quantitative forecasting in Tokai District. *J. Meteorol. Res. (Tokyo)* **4,** 981–984.
50. Yamashita, H. (1952). A method of weather forecasting by meteorological change near tropopause. *J. Meteorol. Res. (Tokyo)* **4,** 179–185.
51. Yule, G. U., and Kendall, M. G. (1940). Introduction to the Theory of Statistics. J. L. Lippincott Co., Edinburgh, 570 pp.
52. Ogawara, M., Ozawa, T., and Tomatsu, K. (1951). Diagrams of confidence limits of occurrence probability. *Meteorol. Soc. Jap. J. Ser. 2.* **29,** 181–193.
53. Holloway, J. L., and Woodbury, M. A. (1953). The Meteorological Statistics Project of the University of Pennsylvania (Abstr.). *Bull. Am. Meteorol. Soc.* **34,** 423.

54. Darling, D. A. (1945). Statistical weather forecasting Report No. 13 (U.S.A.F. contract). Meteorology Dept. California Institute of Technology, Pasadena, California, 26 pp.
55. Grytøyr, Elias (1950). A method of selecting analogous synoptic situations and the use of past synoptic situations in forecasting for East-Norway. *Geofis. Publik.* **17,** 28 pp.
56. Harrison, H. T. (1951). Results of synoptic tests applied to certain forecasting rules. *Bull. Am. Meteorol. Soc.* **32,** 89–96.
57. Johnson, C. B. (1952). Favorable spring frost conditions in the middle Atlantic States. *Weatherwise* **5,** 38–40.
58. Fawbush, E. J., and Miller, R. C. (1953). The tornado situation of 17 March 1951. *Bull. Am. Meteorol. Soc.* **34,** 139–145.
59. Schmidt, R. C. (1952). An objective aid in forecasting precipitation at Washington, D. C., during the month of May. *Bull. Am. Meteorol. Soc.* **33,** 227–232.
60. Schmidt, R. C. (1951). A method of forecasting occurrence of winter precipitation two days in advance. *Monthly Weather Rev.* **79,** 81–95.
61. George, J. J., and collaborators (1953). Forecasting relationships between upper level flow and surface meteorological processes. *Geophys. Res. Pap. U.S.A.F.* No. **23,** 186 pp.
62. Riehl, H., Norquest, K. S., and Sugg, A. L. (1952). A quantitative method for the prediction of rainfall patterns. *J. Meteorol.* **9,** 291–298.
63. Teweles, S. (1953). A test of the relation between precipitation and synoptic patterns at 200 and 300 millibars. *J. Meteorol.* **10,** 450–456.
64. Fletcher, R. D. (1951). Hydrometeorology in the United States, *in* Compendium of Meteorology. American Meteorological Society, Boston, pp. 1033–1047.
65. Linsley, R. K. (1951). The hydrologic cycle and its relation to meteorology-river forecasting, *in* Compendium of Meteorology. American Meteorological Society, Boston, pp. 1048–1054.
66. Landsberg, H. E., and Jacobs, W. C. (1951). Applied climatology, *in* Compendium of Meteorology. American Meteorological Society, Boston, pp. 976–992.
67. Brier, G. W. (1946). A study of quantitative precipitation forecasting in the TVA Basin. U. S. Weather Bureau Research Paper No. 26, Washington, D. C., 40 pp.
68. U. S. Air Weather Service (1953). Forecasting study for Turner Air Force Base, Washington, D. C. Technical Report No. 105–95, 55 pp.
69. Means, L. L. (1952). On thunderstorm forecasting in the Central United States. *Monthly Weather Rev.* **80,** 165–187.
70. Klein, M., and Yifrach, Y. (1952). Quantitative forecasting of the daily minimum temperature during the winter months for Lydda Airport, Israel. *Meteorol. Service Ser. A, Meteorol. Notes* No. **5,** 19 pp.
71. Gringorten, I. I., and Press, H. (1948). A meteorological measure of maximum gust velocities in clouds. *Nat. Adv. Com. Aeronaut. Notes* No. **1569,** 20 pp.
72. Thompson, J. C. (1952). On the operational deficiencies in categorical weather forecasts. *Bull. Am. Meteorol. Soc.* **33,** 223–226.
73. Gringorten, I. I. (1952). An objective aid to local forecasting. *Bull. Am. Meteorol. Soc.* **33,** 202–207.
74. Armstrong, H. (1953). Forecasting tornadoes in Georgia. *Monthly Weather Rev.* **81,** 290–298.
75. Krown, L. (1953). Detailed radiosonde analysis in local 12-hour forecasting of precipitation at Lydda Airport, Israel. *Meteorol. Service Ser. A, Meteorol. Notes* No. **8.**

76. Gilbert, C. G. (1953). An aid for forecasting the minimum temperature at Denver, Colorado. *Monthly Weather Rev.* **81,** 233–245.
77. Krick, I. P. (1937). Forecasting the dissipation of fog and stratus clouds. *J. Aero. Sci.* **4,** 366–371.
78. Abercromby, R. (1885). Principles of forecasting by means of weather charts. H. M. Stationery Office, London, 123 pp.
79. Washburn, H. (1948). An aid to the forecasting of maximum temperatures. Meteorol. Div., Dept. of Transport, Canada. CIR-1469, TEC-46.
80. Zenone, E. (1952). L'influenza de ciclone de Genova sul tempo del Canton Ticino. *Geofis. Pura. Appl.* **21,** 32–38.
81. Vernon, E. M. (1952). A numerical approach to rainfall forecasting for San Francisco (Abstr.). *Bull. Am. Meteorol. Soc.* **33,** 127.
82. Woodbridge, D. D., and Decker, F. W. (1953). Weather-typing applied to summer frontal rains in Oregon. *Bull. Am. Meteorol. Soc.* **34,** 29–37.
83. Franssila, M. (1948). On the forecasting of frost with Ångström's formula. *Geophysica* (*Helsinki*) **3,** 102–113.
84. Elliott, R. D., Olson, C. A., and Strauss, M. A. (1949). Analysis of replies to a questionnaire on forecast aids. *Bull. Am. Meteorol. Soc.* **30,** 314–318.
85. Whiting, R. M. (1948). A method of forecasting the degree of turbulence in fronts and line squalls. Eastern Air Lines, Inc., Meteorol. Dept., Atlanta, Ga., 22 pp.
86. Benton, G. S., Estoque, M. A., and Dominitz, J. (1953). An evaluation of the water vapor balance of the North American continent. Scientific Report No. 1 (USAF Contract No. AF19(122)-365). Johns Hopkins University, Baltimore, 101 pp.
87. Finkelstein, J. (1952). A test of Brunt's formula for predicting minimum temperatures. *New Zealand Meteorol. Services, Tech. Notes* No. **93,** 7 pp.
88. Kussman, A. S. (1948). A statistical method of forecasting maximum temperatures for New York City. Thesis (M.S.), New York University, College of Engineering.
89. Richards, T. L. (1954). An approach to forecasting snowfall amounts. Meteorol. Div., Dept. of Transport, Canada. CIR-2421, TEC-177.
90. Showalter, A. K. (1950). An approach to quantitative forecast of precipitation (III). *Bull. Am. Meteorol. Soc.* **31,** 23–25.
91. Schroeder, M. J. (1950). Forecasting humidities along the north shore of Lake Superior. U. S. Weather Bureau, Short-Range Forecast Development Section, Washington, D. C., 10 pp.
92. Legg, E. M. (1952). Notes on forecasting the minimum relative humidity in Western Washington, U. S. Weather Bureau, Pomona, Calif., 7 pp.
93. Gifford, F. Jr. (1950). Forecasting the heights of formation of stratus clouds. *Bull. Am. Meteorol. Soc.* **31,** 31–37.
94. Crutcher, H. L., Hunter, J. C., Sanders, R. A., and Price, S. (1950). Forecasting the heights of cumulus cloud-tops on the Washington-Bermuda airways route. *Bull. Am. Meteorol. Soc.* **31,** 1–7.
95. Goldman, L. (1951). On forecasting ceiling lowering during continuous rain. *Monthly Weather Rev.* **79,** 133–142.
96. Vernon, E. M. (1952). A numerical approach to rainfall forecasting for San Francisco (Abstr.). *Bull. Am. Meteorol. Soc.* **33,** 123.
97. MacDonald, T. H. (1947). New York snow forecasting problem. Preliminary Report, U. S. Weather Bureau Short-Range Forecast Research Unit, Washington, D. C., 18 pp.

98. Blue Hill Meteorological Observatory (1949). Mt. Washington and Blue Hill data aids to snowstorm forecasting for Boston. Contract Cwb-8099, Final Report, 1948–49 on Snowstorm Phase, 48 pp.
99. Blue Hill Meteorological Observatory (1950). Mt. Washington and Blue Hill aids to snowstorm forecasting for Boston. Contract Cwb-8120, Final Report, 1949–50 on Snowstorm Phase, Typescript, 53 pp.
100. Murray, R. (1952). Rain and snow in relation to the 1000-700-mb and 1000-500-mb thicknesses and the freezing level. *Meteorol. Mag.* **81,** 5.
101. Saunders, W. E. (1950). Method of forecasting the temperature of fog formation. *Meteorol. Mag.* **79,** 213–219.
102. Ibaragi, T. (1953). A note on the forecast of fog occurrence. *J. Meteorol. Res. (Tokyo)* **5,** 185–188.
103. Brablec, J. (1950). Předpověd nočnich mrazu podle Sinjavskiho. *Meteorol. Zpravy (Prague)* **4,** 35–36.
104. Beers, F. D. (1948). A preliminary study of the problem of forecasting northerly winds in the Sacramento Valley during the fire weather season. U. S. Weather Bureau, Washington, D. C., 5 pp.
105. Huntoon, J. K. (1950). Forecasting strong northwesterly winds in the Los Angeles Basin with the aid of the 500 millibar chart. U. S. Weather Bureau, Airport Station, Los Angeles, Calif., 5 pp.
106. U. S. Air Weather Service (1952). Forecasting jet aircraft condensation trails. Air Weather Service Manual 105–100, Washington, D. C., 17 pp.
107. U. S. Air Weather Service (1951). The Cirrus-forecasting problem. U. S. Air Weather Service Tech. Report No. 105–81, 11 pp.
108. U. S. Weather Bureau, Short-Range Forecast Development Section (1951). A meteorological forecasting investigation for improving the prediction of airport acceptance rates. U. S. Weather Bureau, Washington, D. C., Final Reports, Parts I and II, 91 pp.
109. Bristor, C. L. (1952). Relating prognostic charts to winter precipitation at Sioux City, Iowa. U. S. Weather Bureau, Washington, D. C.
110. U. S. Air Weather Service (1953). Some techniques for deriving objective forecast aids and methods. U.S.A.F. Air Weather Service Manual 105–40, 31 pp.
111. Brier, G. W., and Allen, R. A. (1951). Verification of weather forecasts, *in* Compendium of Meteorology. American Meteorological Society, Boston, pp. 841–848.
112. Gringorten, I. I. (1951). The verification and scoring of weather forecasts. *J. Am. Statistical Assoc.* **46,** 279–296.
113. Crossley, A. F. (1953). Measures of success in forecasting. Air Ministry, Meteorological Research Committee, M.R.P. 788, S.C. II/138, 12 pp.
114. Vernon, E. M. (1953). A new concept of skill score for rating quantitative forecasts. *Monthly Weather Rev.* **81,** 326–329.
115. Leight, W. G. (1953). The use of probability statements in extended forecasting. *Monthly Weather Rev.* **81,** 349–356.
116. Gringorten, I. I. (1954). Test of significance in a verification program. (Unpublished.)
117. Jorgensen, D. L. (1953). Estimating precipitation at San Francisco from concurrent meteorological variables. *Monthly Weather Rev.* **81,** 101–110.
118. Köppen, W. (1884). Eine rationelle Methode zur Prüfung der Wetterprognosen. *Meteorol. Z.* **1,** 397–404.
119. Sawyer, J. S. (1952). A study of the rainfall of two synoptic situations. *Quart J. Roy. Meteorol. Soc.* **78,** 231–246.

120. Oliver, V. J., and Oliver, M. B. (1951). Meteorological analysis in the Middle Latitudes, *in* Compendium of Meteorology. American Meteorological Society, Boston, pp. 715–727.

121. Chambers, E. (1952). "Comets" and weather. *Weather* **7,** 67–72, 111–116.

122. Lederer, J. (1953). Some relationships between weather and air safety. *Bull. Am. Meteorol. Soc.* **34,** 339–350.

123. Spar, J. (1953). A suggested technique for quantitative precipitation forecasting. *Monthly Weather Rev.* **81,** 217–221.

124. Fawbush, E. J., and Miller, R. C. (1954). A basis for forecasting peak wind gusts in non-frontal thunderstorms. *Bull. Am. Meteorol. Soc.* **35,** 14–19.

Wind Generated Gravity Waves

WILLARD J. PIERSON, Jr.

Department of Meteorology and Oceanography, College of Engineering, New York University, New York, N.Y.

1. Introduction

1.1. History of Recent Developments

In 1949 a conference was held at the New York Academy of Sciences on the subject of ocean surface waves. Theoretical and practical papers of great importance were presented. M. A. Mason [1], then of the Beach Erosion Board, Corps of Engineers, gave a paper in which he said:

> Our chief objective remains to be reached. In spite of familiarity with the problem for several thousand years, we still find ourselves in the indefensible position of attempting to study a natural phenomenon which we have not adequately defined. Borrowing from Thorade [2], we may repeat in 1948 what he wrote in 1931—"No adequate results of observation are available

> with respect to form, orbital path, and energy, but they are sufficient to shake our confidence in theory . . . Complete agreement between theory and observation is seldom found, and where it is found, it seems suspicious."
>
> What is an ocean surface wave? I believe I can truthfully say that we cannot define an ocean surface wave from observation. You and I have watched waves on the Pacific Coast that seemed to be regular and, to some extent, uniform, but even casual observation showed that consecutive waves varied widely in period, form, and estimated height (an observation confirmed by many instrumental measurements of waves on the same coast).

At the time of the conference, no one could have answered the question, "What is an ocean surface wave?" Simple harmonic progressive waves, solitary waves, and individual breaking waves were discussed. Actual wind generated gravity waves were being analyzed for "significant" height and "significant period." Wave spectra were being studied by Deacon [3] and Klebba [4].

In August 1950, a symposium on mathematical statistics and probability was held in California. At this symposium, Rudnick [5] suggested that ocean wave records could be represented by a stationary Gaussian process as a function of time at a fixed point. This was the first hint of an adequate definition of wind generated gravity waves. It was shown that points from a pressure wave record have a normal distribution.

In 1951, a conference on gravity waves was held at the National Bureau of Standards. Many important properties of internal waves, simple harmonic progressive waves, and other types of periodic wave motions were discussed. At this conference, Birkhoff and Kotik [6] again suggested that an ocean wave record could be represented by a stationary Gaussian process, but their study of some actual records led them to be dubious of this hypothesis in that minor nonlinear effects had distorted the probability distribution function so that it was not quite Gaussian.

In 1952, Pierson [7] published a paper which argued that the Gaussian hypothesis should be accepted as describing a wave record despite minor nonlinear effects. A short-crested Gaussian sea surface was suggested as a model, and problems in the analysis of wave records and the propagation of waves were treated. In this paper, these results will be brought up to date and combined with the results of many other workers. A fairly complete theory of wind generated gravity waves will be developed.

1.2. Purpose and Procedure

The purpose of this article is to describe new theoretical developments which give a more complete and more accurate description of the analysis, generation, propagation, and refraction of wind generated gravity waves.

A special path of development through hydrodynamic theory permits a link with time series theory which yields a realistic and practical description of actual ocean waves. This description explains the irregularity of these waves, it describes their short crestedness, and it illustrates the change from sea waves to swell waves. Because of the essential irregularity of actual wind generated gravity waves, the finer developments and the refinements of nonlinear theory have to be sacrificed. Linear theories must be used in order to employ the principle of superposition to the best advantage. The irregularity and the marked variation in height from wave to wave in an actual wave record must be explained before minor effects of nonlinearity can be taken into consideration.

The plan of this article is: (1) to outline the many different kinds of inquiry pursued in classical hydrodynamics with reference to waves; (2) to show that simple harmonic progressive waves must be selected as the link with time series theory; (3) to show how Fourier integral theory leads more generally to time series theory; and, finally, (4) to show how a model of wind generated gravity waves can be constructed which will explain many of the features of actual waves. As mentioned above, nonlinear features cannot be treated at the present time, but recent remarks by J. W. Tukey in a paper by St. Denis and Pierson [8] suggest that at a later date, if needed, refinements which will take into account nonlinear effects can be introduced in this theory.

The results presented will show that the questions asked by Mason [1] in 1949 have essentially been answered. An accurate and nearly complete description of ocean waves as they actually are has been obtained.

2. Outline of Theory

2.1. Schematic Diagram

Figure 1 shows a schematic diagram of the elements needed to construct a theory of wind generated gravity waves. Developments in classical hydrodynamics, Fourier integral theory, generalized harmonic analysis, time series theory, and statistics need to be combined to produce this theory of wind generated gravity waves. The main lines of development which are sketched show the path which must be followed. Parts of classical hydrodynamic theory have been omitted, and the diagram given in Fig. 1 is not meant to be complete. It does, however, show what needs to be known here. The main path of development is shown by the heavy lines in the diagram. For the application of classical hydrodynamic theory to oceanography, and for further information on the history of the development of wave theory, see "Dynamical Oceanography" by Proudman [9].

THEORETICAL AND APPLIED MATHEMATICS

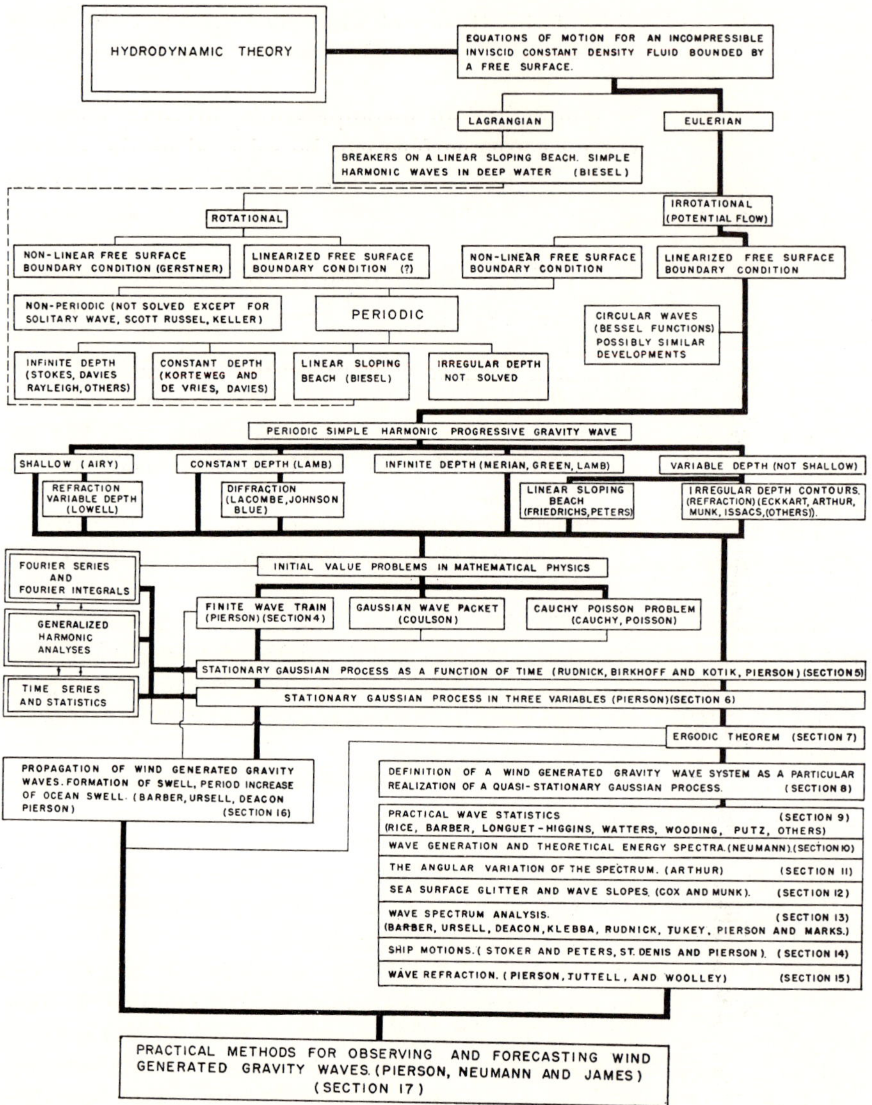

FIG. 1. Schematic diagram of theoretical developments which lead to a description of wind generated gravity waves.

2.2. Classical Hydrodynamics

In this discussion, Fig. 1 will be used to show the path which is followed in the derivation of this theory of wind generated gravity waves. Only part of the field of hydrodynamic theory, which includes many more theoretical results, is shown here.

The equations of motion for an incompressible, inviscid, constant-density fluid bounded by a free surface can be selected as the starting point. These equations may have one of two possible forms. They can be expressed in a Lagrangian reference system or in an Eulerian reference system.

The work of Biesel [10] in a Lagrangian reference system explains breakers on a linear sloping beach in one of the most realistic ways ever achieved. The figures which illustrate his article [10] are the most realistic representation of a breaker ever presented. Certainly the line of attack opened up by this paper in the use of the Lagrangian system needs to be developed and studied further in connection with the rest of this particular study. Not much work has been done using the Lagrangian system up to the present.

The main line of attack selects the Eulerian equations. The choice of irrotational (potential) flow as opposed to rotational flow must then be made.

An exception is the Gerstner wave as described by Lamb [11] for flow which is not irrotational that occurs in the theoretical development at this point. This wave appears to be an anomaly in the main line of the theory, and there does not seem to be much practical utility in further investigation of it.

Potential flow in an Eulerian reference system has thus been chosen to represent the fluid motion. The free-surface boundary condition then presents the major mathematical difficulty in the development of the theory of possible resulting fluid motions. There are two choices. The free surface boundary condition can be kept in its nonlinear form, or it can be linearized at this point in the development.

If the nonlinear free-surface boundary condition is chosen, this severely restricts the types of problems which can be handled. In fact, the general case of nonperiodic solutions cannot be treated, and the only path of development which then results occurs when periodic solutions are chosen along with this more general nonlinear free-surface boundary condition.

Solutions which produce periodic waves and which are refinements of the theory of a simple harmonic progressive wave, as far as the wave profile and wave properties are concerned, have been developed by many workers. For infinite depth, the work of Stokes [12] as carried out in 1847 is important. For infinite depth and constant depth, the recent work of Davies [13, 14] has practically completed this aspect of the theory of gravity waves.

For variable depth, although the boundary condition at the bottom, which requires that the fluid velocity components normal to the bottom are zero, is linear, the problem becomes difficult if the free-surface boundary condition is nonlinear, and it has not been completely solved. The problem has not been attempted at all for arbitrary bottom configurations. However, Biesel's method of attack, which employs a Lagrangian system for the equations of motion, permits successively higher and successively more accurate approximations. Thus in a sense this particular problem has also been solved to a high degree of accuracy for a linear sloping beach.

The difficulty, however, in any of these problems, as far as applying them to actual conditions, is that the sum of two solutions is not a solution to a new problem. Stated another way, a non-linear free-surface wave profile with a period of ten seconds cannot be combined with a nonlinear free-surface wave profile with a period of eleven seconds to produce a system of waves which satisfies the conditions which are imposed at the free surface. All of the very important and extremely useful techniques developed in Fourier integral theory (and in what will follow) cannot be applied.

The main line of development, therefore, must proceed with the assumption of a linearized free-surface boundary condition. When this is done, the distinction between periodic and nonperiodic solutions becomes less important because the periodic solutions can be used to develop any desired nonperiodic solution. Thus the main line of development proceeds immediately to the study of a simple harmonic progressive wave with infinitely long, straight, parallel crests and with one discrete spectral period.

At this point, a possible side development can be mentioned. An alternate solution of the equations which result from the assumptions described is one involving Bessel function waves which radiate from a point source in ever-expanding circles. It may be possible, beginning with such waves, to develop a complete parallel theory for such waves which will help to explain the phenomenon of the propagation of waves from hurricanes and the generation of wind driven gravity waves in hurricanes. However, this particular branch of the theory has not been extensively studied as of the time of this paper.

If the free-surface boundary condition is linearized, a solution is the simple harmonic progressive wave. This elemental simple harmonic progressive wave was one of the first developments of classical hydrodynamics. Shallow water theory as developed by Airy [15] for nondispersive waves, where the wave crest speed is simply given by $\sqrt{gh}$ for constant depth, was one of the earliest developments. The theory of simple harmonic waves in deep water and in water of constant depth as presented by Lamb [11] and as derived by Merian [16], Green [17], and others was developed in the years from 1820 to 1850.

More recent advances in this particular line of development have occurred. In the theory of the propagation of waves in shallow water, the work of Lowell [18] has given a complete mathematical formulation to the problem of the refraction of shallow water gravity waves. For constant depth, the diffraction of simple harmonic progressive waves has been studied by Lacombe [19] and Johnson and Blue [20].

The last stage of wave refraction may, in some cases, produce waves with a profile like that of the solitary wave. The work of Keller [21] has formalized the representation of the solitary wave.

The problem of variable depth has most recently yielded to successful mathematical techniques. The problem of gravity waves approaching a linear sloping beach has been solved by Friedrichs [22] and Peters [23]. The problem solved by Friedrichs was that of a simple harmonic progressive wave in deep water approaching perpendicular to a beach with the depth increasing linearly away from the beach. Thus the problem was that of finding a potential function in a pie-shaped wedge which satisfied the appropriate free-surface boundary conditions and the appropriate condition at the slanting bottom, and which, at the same time, had the properties of the potential function of a simple harmonic progressive wave after the water had become sufficiently deep. Peters extended the problem to one which considered wave crests which were not parallel to the depth contours in deep water and showed the shape of the arcs which the crests bent into as they approached the beach. Both of these solutions are exact, and the potential function for the wave motion is exhibited explicitly in the problem. It is, however, still a linear solution, and the form of the wave crests near the beach becomes very strongly distorted since there is a singularity in the potential function at the beach. The waves do not break as they do in Biesel's solution.

For irregular depth contours, even the linearized problem has not been solved exactly. The most rigorous approach to the solution of the problem of the propagation of waves over a bottom of variable depth has been given by Eckart [24]. After a few well-justified simplifying assumptions, equations are derived which explain the refraction of simple harmonic progressive waves.

Practical methods for the graphical construction of refraction diagrams have been studied and developed over the past few years at the University of California. An early work on the subject was that of Johnson, O'Brien, and Isaacs [25] which developed the concept of the wave ray and applied principles for the construction of this ray to the problem instead of employing a construction method which used the wave crests. This procedure eliminated many of the inaccuracies which resulted from the wave crest method and demonstrated the existence, for example, of caustics as discussed by Pierson [26]. The most recent development along these lines is that of Arthur, Munk, and Isaacs [27] which refines the method for constructing the wave rays and eliminates sources of error which resulted from approximations used in the previous methods. This technique is undoubtedly the most practical one available for the solution of complex problems in a practical case where wave refraction must be studied.

The basic assumption of wave refraction is that the dimensions of the refracting object are large compared to the wavelength of the waves involved. Just when this assumption begins to fail is unknown at the present state of the development of the theory.

The present status of the theory of the refraction and diffraction of simple harmonic gravity waves can be compared with an analogous situation in the theory of optics. Refraction theory is essentially the same as the theory of geometrical optics in the theory of light. Diffraction theory is essentially a collection of rather special solutions to types of diffraction problems. Geometrical optics has been extended to physical optics by a closer study of the appropriate equations, and the work of Luneberg [28] shows in part how this can be done. One possible extension of gravity wave theory, therefore, lies in a study of these works and in an attempt to extend current theory to a theory which would successfully include both. The result could lead to equations quite analogous to those of physical optics. They would give information on the point at which the assumptions employed in current-wave refraction theory become invalid.

2.3. Fourier Integral Theory

A brief review of what has been given so far, as summarized in Fig. 1, shows that the main line of approach has yielded many valuable studies which are based upon the properties of the simple harmonic progressive gravity wave. Since these results are based on linear equations they can be combined to yield information on actual wind generated gravity waves at a later stage of development. This very important property of linearity is most useful. It provides only an approximation because the true conditions are non-linear, but it permits a representation of the waves which will prove to be more realistic than any other possible representation.

The Fourier integral permits the representation of a function which satisfies certain mathematical conditions by an integral which involves trigonometric terms. The function to be represented must die down at plus or minus infinity sufficiently rapidly (or vanish) to be acceptable.

In the classical wave equation of mathematical physics, the wave crest speed is not a function of the frequency and the waves are non-dispersive. In this classical case, wave propagation depends only on the characteristic curves of the solutions. However, in gravity wave theory, each spectral wavelength must be assigned an appropriate spectral period, or conversely, each spectral period must be assigned an appropriate wavelength. This, in turn, leads to the concept of group velocity. Thus the classical initial-value problem of mathematical physics which is used as an analogue must be interpreted somewhat differently in order to solve problems in gravity wave propagation.

The classical problem which first treated the propagation of gravity waves is the one solved by Cauchy and Poisson* as given (in a readily available reference) by Lamb [11]. This problem formulates the general initial-value problem. It shows that if the free surface as a function of distance is given at a time $t = 0$, and if the derivative of the free surface with respect to time as a function of distance is given at the time $t = 0$, an integral can be formulated which will describe the propagation of the waves at later times.

As an illustration of their theory, Cauchy and Poisson solved the problem of the propagation of waves generated by a concentrated point source in infinitely deep water. This source is an infinitely high, infinitesimally wide (and infinitely long in the y direction) column of water which begins to fall into the sea surface at the time $t = 0$ and which produces waves which propagate away from the resulting disturbance. The Cauchy-Poisson problem was the first to illustrate the effects of group velocity and to formulate the concept of group velocity for gravity waves. However, it is not too realistic because it does not lend itself readily to the study of actual waves. The reason for this is that the spectrum of the motion contains all frequencies with equal intensity. Actual gravity waves, as will be shown later, do not contain low frequencies (high periods) and thus the period increase of ocean swell has to be explained in a somewhat different way.

Also representative of the type of problem that can be solved using these techniques is the problem discussed by Coulson [29], which is called the Gaussian wave packet. This problem, again, is not too realistic

* See also p. 400 of Proudman [9].

because of the assumption which is made in evaluating an integral in order to obtain the solution of the problem. The assumption is the stationary phase approximation. It results in a solution which does not reduce to the original initial values, and thus the behavior of the system a short distance away from the source is not realistic. There are many other problems in the literature which have been formulated and solved by similar techniques. As far as can be determined, all of these solutions which have been presented employ the stationary phase approximation, and therefore they are not as realistic as they could be.

Most of the problems which have been treated have been studied in two variables, x and t. Fuchs [30] studied the problem of the propagation of short crested waves out of an area on the sea surface. The stationary phase approximation was used, and the result is rather surprising in that the highest waves received at a distant point are found to occur at an angle to the direction in which the apparent waves are traveling in the source region. As will be shown later, the reason for this result is that the short crested model which is used is really just the sum of two long crested wave systems traveling at an angle to each other.

It is possible to avoid the stationary phase approximation in the study of certain models. One example is given by Pierson [31], and in this article an extension will be given which permits the solution of the problem of the propagation of a finite train of simple harmonic waves restricted to a rectangular area of the sea surface at the source.

The study of Fourier integral methods for the representation of the sea surface and of wave records will never yield a complete explanation of wind generated gravity waves. The result which will be given in the section of this article which treats Fourier integral problems will not be realistic since the result will not be like those obtained from the study of actual ocean waves. However, it will form part of the chain of the theory which will finally explain the formation of swell and the period increase of ocean swell. The missing link is, of course, the application of generalized harmonic analysis and the principles involved in the study of stationary Gaussian processes.

2.4. Generalized Harmonic Analysis and Time Series

Fourier integral theory is very useful and can be used to solve many problems in mathematical physics. Its application could solve this problem of wave propagation if the initial values could be determined over a huge area, say 500 miles on a side. When one considers that, in a wind generated sea, there are from ten to twenty waves per mile and that their heights are completely irregular, the problem becomes one which is far too complex to handle by Fourier integral methods.

Escape from this difficulty is possible by means of generalized harmonic analysis, time series theory, and statistics. New mathematical developments of communication theory then become applicable to the description of waves. Only one particular aspect of generalized harmonic analysis and time series theory is needed. It is the theory of stationary Gaussian processes as described by Lévy [32], Rice [33], Tukey [34], Wiener [35], and others. When this theory of stationary Gaussian processes is combined with the principles of the simple harmonic progressive gravity wave to yield a three-dimensional stationary Gaussian process, the result is a most realistic representation of wind generated gravity waves.

2.5. Final Results

The main line of development thus goes from classical hydrodynamics by means of simple harmonic progressive waves through all appropriate linear developments in classical hydrodynamics to the theory of the stationary Gaussian process in three variables. The result is an answer to the question: "What is an ocean surface wave?"

With the answer to this question, it then becomes possible for the geophysicist to observe and study actual waves. The bottom part of Fig. 1 shows the advances which have been made and gives the section in which they will be treated in this paper.

An abstract theory is of little value without concrete results. The work of Neumann [36–39], begun in 1950, in deriving the spectrum of the waves, is a major contribution to the theory of wind generated gravity waves because it links theoretical ideas with practical results. The theory of wave generation developed by Neumann makes it possible to obtain practical results in forecasting and observing actual waves.

3. The Simple Harmonic Progressive Wave

3.1. Equations for the Formulation

The derivation of the properties of a simple harmonic progressive wave is a part of classical hydrodynamics. The procedure follows the lines shown in Fig. 1. After the various assumptions which were listed have been made, the result is the potential function given by*

$$\phi_{xx} + \phi_{yy} + \phi_{zz} = 0 \tag{3.1}$$

The linearized free-surface boundary conditions at $z = 0$ are given by

$$\eta = \frac{1}{g}\,\phi_t \qquad \text{at} \qquad z = 0 \tag{3.2}$$

$$\eta_t = -\phi_z \qquad \text{at} \qquad z = 0 \tag{3.3}$$

* Subscripts denote partial differentiation.

The boundary condition at the bottom is given by

$$\phi_z = 0 \quad \text{at} \quad z = -h \tag{3.4}$$

The potential function which satisfies the above conditions is given by

$$\phi = \frac{-AgT}{2\pi \cosh \frac{2\pi h}{L}} \sin \left(\frac{2\pi}{L} (x \cos \theta + y \sin \theta) - \frac{2\pi t}{T} + \epsilon \right) \cdot \cosh \left[\frac{2\pi}{L} (z + h) \right] \tag{3.5}$$

This solution is the most general one possible for a simple harmonic progressive wave in water of constant depth. It should be noted that the simple harmonic progressive wave which is immediately obtainable from equation (3.2) upon substitution of the derivative of (3.5) with respect to time has a free-surface profile which can travel in any direction as determined by the angle θ, which is measured with respect to the positive x-axis.

When this potential function is substituted into equations (3.2) and (3.3), the result is that equation (3.6) must also be satisfied. In infinitely deep water, the wave speed is given by $gT/2\pi$, and in shallow water, the wave speed is given by $\sqrt{gh}$.

$$C^2 = \frac{gL}{2\pi} \tanh \frac{2\pi h}{L} = \frac{L^2}{T^2} \tag{3.6}$$

Equation (3.6) is a very difficult equation to handle in the derivation of Fourier integral problems because it is a transcendental equation. In deep water, equation (3.6) simplifies, and very nearly all Fourier integral problems are solved for the deep water propagation case.

For certain applications in wave propagation it is necessary to study the group velocity of the waves in shallow water and the result of transition from deep water, where the group velocity is half of the wave crest speed, to shallow water, where it is independent of the period of the wave. Under these conditions, the solutions of Fourier integral problems become difficult. No general solutions of initial value problems in this transition zone are available.

The potential function given by equation (3.5), or the free surface which would result from it, involves 2π over the wavelength and 2π over the period, where the wavelength and period are related by equation (3.6). The wavelength can be expressed as a tunction of the period and the depth or, as will be more convenient in this paper, as a function of the wave frequency and the depth. The wave frequency, μ, is given by

$2\pi/T$, and equation (3.7) can be written down

$$C = \frac{L}{T} = \frac{L\mu}{2\pi} \tag{3.7}$$

With the aid of this equation, (3.6) can be put into the form

$$\frac{2\pi}{L} = \frac{\mu^2}{g} \coth\left(\frac{h2\pi}{L}\right) \tag{3.8}$$

Equation (3.8) could be repeated a second time, and the result would be two equations involving $2\pi/L$ and μ^2/g. The $2\pi/L$ occurs on the left side of each equation and as part of the argument of the hyperbolic cotangent. When this second copy of equation (3.8) is then substituted into the right side of equation (3.8), the result is equation (3.9); and equation (3.8) can then be substituted into the far right of equation (3.9) to yield equation (3.10)

$$\frac{2\pi}{L} = \frac{\mu^2}{g} \coth\left(\frac{h\mu^2}{g} \coth\left(\frac{h2\pi}{L}\right)\right) \tag{3.9}$$

$$\frac{2\pi}{L} = \frac{\mu^2}{g} \coth\left(\frac{h\mu^2}{g} \coth\left(\frac{h\mu^2}{g} \coth\left(\frac{h\mu^2}{g} \cdots\right)\right)\right) \tag{3.10}$$

$$\frac{2\pi}{L} = \frac{\mu^2}{g} \operatorname{Itcoth}\left(\frac{h\mu^2}{g}\right) = \frac{\mu^2}{g} I\left(\frac{\mu^2 h}{g}\right) \tag{3.11}$$

The result as indicated by equation (3.10) can be repeated indefinitely, and the right side of the equation becomes essentially a function of μ^2/g and h. It is thus independent of the shallow-water wavelength, L. This is a rather surprising result as shown by Pierson [7], but the factor $h\mu^2/g$ can be evaluated for a given depth and a given wave frequency, and its hyperbolic cotangent can be computed. After this hyperbolic cotangent is multiplied by the factor $h\mu^2/g$, the hyperbolic cotangent of the product can be computed. This process can be repeated, or iterated, indefinitely, and the result will converge to the value of $2\pi/L$. For fairly typical values of $h\mu^2/g$, seven or eight iterations will yield a result which gives the value of $2\pi/L$ to three or four significant figures.

Thus $2\pi/L$ can be written as equation (3.11) where the new symbol stands for the interated hyperbolic cotangent of μ^2h/g. This particular notation is extremely useful in studying certain refraction effects because it exhibits the dependence of the wavelength on the frequency and the depth explicitly. It is also convenient to be able to express the potential function and the equation for the free surface in terms of the depth of the water and the frequency of the wave.

3.2. Deep Water

The deep-water case of this problem is by far the simplest for application to propagation problems. Under these conditions the above equations simplify very much, and the free surface is given by

$$\eta(x,y,t) = A \cos \left[\frac{\mu^2}{g} (x \cos \theta + y \sin \theta) - \mu t + \epsilon \right] \tag{3.12}$$

Equation (3.12) involves four parameters. They are: (1) A, the amplitude of the waves, (2) μ, the wave frequency, equal to $2\pi/T$, (3) θ, the direction toward which the wave is traveling, as measured in a counterclockwise direction with respect to the x-axis, and (4) ϵ the phase of the waves at $x = y = t = 0$. Complete freedom of choice in the four parameters is needed in order to describe waves in the most general way. The basic representation of wind generated gravity waves in deep water will be constructed from a linear superposition of such elemental simple harmonic progressive waves.

3.3. Added Comments

The derivation of this very simple case has been repeated here. The derivation for any of the other cases discussed in the outline which can be obtained from the assumption of a linearized free-surface boundary condition could be employed at this point to obtain the behavior of a simple harmonic progressive wave under the conditions studied. All of the results which were outlined above can therefore be used at a later stage with appropriate modifications to derive other needed properties of wind generated gravity waves.

Such properties as the velocity components at any depth, the sea surface slope and the pressure at any depth can be immediately computed from the derivations given above in the case of the simple harmonic progressive wave. Therefore, although it will not be carried out explicitly for all possible applications, almost any property of actual wind generated gravity waves can be studied if that property is understood in simple harmonic progressive waves.

3.4. Danger of Indiscriminate Use of Above Theory

Wind generated gravity waves are not simple harmonic progressive waves. The usual procedure in many references in the literature is to assign a "significant period" and a "significant height" to the wave given by equation (3.12), or its appropriately modified form for refraction cases or shallow-water cases, and assume that the wind generated waves will have properties deducible from this equation. This is far from the case, as subsequent results will show.

4. A Finite Wave Train in Deep Water

4.1. Formulation

In this section, the solution of an initial value problem will be presented in terms of a slightly modified, but still rigorous, Fourier integral method. The usual procedure is to treat a two-variable problem as a function of x and t, to prescribe $\eta(x)$ at $t = 0$ and $\eta_t(x)$ at $t = 0$, and solve. When the problem is formulated in this way the integrals which result usually require the employment of the stationary phase approximation* in their evaluation. This makes the results obtained partly unrealistic.

Instead, in this problem, a three-variable problem will be treated in which the free surface $\eta(y,t)$ at $x = 0$ is prescribed, and an assumed direction for the propagation of the spectral components is taken. This eliminates the necessity of formulating the derivative of the motion at $x = 0$ and thus simplifies the integrals which need to be evaluated. Certainly, the assumption about the direction of propagation of the various spectral components is a realistic assumption when wind generated gravity waves are being considered since it would seem logical to assume that the waves will travel in the direction of the wind. It will be shown, as this development progresses, that the results are extremely realistic.

4.2. The General Solution

The problem to be treated in general, then, can be stated as follows: given $\eta(y,t)$ at $x = 0$, and given that all *important* spectral components are traveling in the positive x-direction, find $\eta(x,y,t)$. Suppose that the free surface as a function of x, y, and t can be represented by

$$\eta(x,y,t) = \int_{-\pi/2}^{\pi/2} \int_0^\infty a(\mu,\theta) \cos\left(\frac{\mu^2}{g}(x \cos\theta + y \sin\theta) - \mu t\right) d\mu d\theta + \int_{-\pi/2}^{\pi/2} \int_0^\infty b(\mu,\theta) \sin\left(\frac{\mu^2}{g}(x \cos\theta + y \sin\theta) - \mu t\right) d\mu d\theta \tag{4.1}$$

This equation is an integral over μ and θ of a large number of simple harmonic progressive waves where their amplitudes are prescribed by a spectrum which is given as a function of two variables. The spectrum is given as $a(\mu,\theta)$ for the even waves and as $b(\mu,\theta)$ for the odd waves.

In equation (4.1), set $x = 0$. Then $\eta(0,y,t)$ is given. In general, $\eta(0,y,t)$ would be a given function of y and t, and it would not be represented in the integral form given by (4.1). In (4.1) the bounds of inte-

* See Lamb [11], Chapter 9, section 24.

gration over θ automatically require that all spectral components travel in the positive x-direction, and therefore enough conditions have been assigned to determine the subsequent motion at other x since the free surface at $x = 0$ is given. The next step, then, is to determine the spectra, $a(\mu,\theta)$ and $b(\mu,\theta)$, from the knowledge of the free surface as a function of y and t.

To do this, a lemma which is found in Courant [40, see page 321, Vol. 2] is employed. This lemma is stated by

$$\lim_{N\to\infty} \int_{-A}^{A} f(x + x') \frac{\sin Nx'}{x'} dx' = \pi f(x) \tag{4.2}$$

The representation of the free surface with $x = 0$ as given by equation (4.1) is multiplied by $\cos(\mu'^2 \sin \theta' y/g)$ and $\cos \mu' t$ and integrated over y and t from $-\infty$ to $+\infty$. This step is indicated in equation (4.3). When the representation of $\eta(0,y,t)$, obtainable from (4.1), is substituted into (4.3) for $\eta(0,y',t')$, and after the indicated integrations are carried out, and this is followed by the use of the lemma given by equation (4.2), the final result is the right side of equation (4.3)

$$\lim_{\substack{N\to\infty\\ M\to\infty}} \int_{-M}^{M} \int_{-N}^{N} \eta(0,y',t') \cos\left(\frac{(\mu')^2}{g} \sin \theta' \cdot y'\right) \cos \mu' t' \, dy' dt' = \frac{\pi^2 g[a(\mu', \theta') + a(\mu', -\theta')]}{(\mu')^2 \cos \theta'} \tag{4.3}$$

Equation (4.3) gives part of the expression for $a(\mu',\theta')$, since it also involves $a(\mu',-\theta')$. When another term which results from continued analysis is evaluated, the result is a right side given by the right side of (4.3), with a minus sign in front of $a(\mu',-\theta')$. When equation (4.3) and this second equation are combined, the unwanted term can be eliminated, and the final result is equation (4.4) which gives one of the needed spectra to be substituted into equation (4.1)

$$a(\mu',\theta') = \frac{(\mu')^2 \cos \theta'}{2\pi^2 g} \int_{-\infty}^{\infty} \int_{-\infty}^{\infty} \eta(0,y',t') \cdot \cos\left(\frac{(\mu')^2}{g} \sin \theta' y' - \mu' t'\right) dy' dt' \tag{4.4}$$

$$b(\mu',\theta') = \frac{(\mu')^2 \cos \theta'}{2\pi^2 g} \int_{-\infty}^{\infty} \int_{-\infty}^{\infty} \eta(0,y',t') \cdot \sin\left(\frac{(\mu')^2}{g} \sin \theta' y' - \mu' t'\right) dy' dt' \tag{4.5}$$

An exactly similar technique gives $b(\mu',\theta')$. Thus from the free surface as a function of y and t, the needed spectra have been determined and the solution to the initial value problem is thus given by equation (4.1) upon the substitution of equations (4.4) and (4.5) into (4.1), and upon the omission of the primes on μ and θ in equations (4.4) and (4.5).

The complete solution is then a function of x, y, and t. The solution is a superposition of simple harmonic progressive waves with different frequencies and directions. Their in-phase amplitudes are determined from the spectra, $a(\mu.\theta)$ and $b(\mu,\theta)$, which depend on the initial form of the function, $\eta(0,y,t)$.

By similar techniques it is possible to solve an initial value problem given by $\eta(x,y,0)$ under the assumption that the spectral components are traveling in the positive x-direction. Equation (4.1) can be used as a starting point, and the procedures are the same.

4.3. Alternate Formulation

An alternate formulation which is frequently found in the literature consists in representing the free surface as a function of x and y by a doubly periodic Fourier analysis. The elementary waves in the sum, when represented in this way, are still simple harmonic progressive waves with crests at various angles to the positive x-axis. However, it is very difficult to sort out the period effect and the direction effect since the period is determined from the sum of the squares of the coefficients of x and y. This is why the above formulation has been presented. It looks like, and is, exactly what happens when an infinite sum of infinitesimally high simple harmonic progressive waves with different frequencies and directions are combined in order to produce a disturbance. Since the component waves propagate with a group velocity which is determined by μ and travel in the direction determined by θ, this representation leads to a better understanding of what is actually occurring in the integral.

4.4. A Special Solution, the Finite Wave Train

Suppose that at the line $x = 0$ in the x,y plane, one observes the conditions given by equation (4.6). Along this line ($x = 0$) as a function of y, a simple sinusoidal oscillation is observed for a total length of D_W seconds in time beginning $D_W/2$ seconds before zero time and ceasing at $D_W/2$ seconds after zero time. As a function of y along this line the disturbance is observed only if y lies between $W_s/2$ and $-W_s/2$. The disturbance, $\eta(y,t)$, is identically zero otherwise at the line $x = 0$. All this is stated in

$$(4.6) \qquad \eta(y,t) = \begin{cases} -A \sin \dfrac{2\pi t}{T} & \text{if } -\dfrac{W_s}{2} < y < \dfrac{W_s}{2} \\ \text{and} & \text{if } -\dfrac{D_w}{2} < t < \dfrac{D_w}{2} \\ 0 & \text{otherwise} \end{cases}$$

The free surface given in equation (4.6) is odd in time and even in y, and therefore the spectrum is easily determined. It is given by evaluating the integral in equation (4.7). When this integration is carried out the result is equation (4.8)

(4.7)

$$b(\mu,\theta) = \frac{\mu^2 \cos \theta}{2\pi^2 g} \int_{-D_w/2}^{D_w/2} \int_{-W_s/2}^{W_s/2} -A \sin \frac{2\pi t}{T} \cos \left(\frac{\mu^2}{g} \sin \theta \cdot y\right) \sin \mu t dy dt$$

$$(4.8) \quad b(\mu,\theta) = \frac{A \cot \theta}{\pi^2} \left[\frac{\sin\left(\left(\mu - \frac{2\pi}{T}\right)\frac{D_w}{2}\right)}{\left(\mu - \frac{2\pi}{T}\right)} - \frac{\sin\left(\left(\mu + \frac{2\pi}{T}\right)\frac{D_w}{2}\right)}{\left(\mu + \frac{2\pi}{T}\right)} \right] \cdot \left[\sin\left(\frac{\mu^2}{g} \sin \theta \frac{W_s}{2}\right)\right]$$

Since the spectrum has been determined, it can be substituted into equation (4.1), and the result is

$$(4.9) \quad \eta(x,y,t) = \int_{-\pi/2}^{\pi/2} \int_{-\infty}^{\infty} \frac{A \cot \theta}{\pi^2} \left(\frac{\sin\left(\left(\mu - \frac{2\pi}{T}\right)\frac{D_w}{2}\right)}{\left(\mu - \frac{2\pi}{T}\right)} \right) \cdot \left(\sin\left(\frac{\mu^2}{g} \sin \theta \frac{W_s}{2}\right)\right) \cdot \sin \left(\frac{\mu^2}{g} (x \cos \theta + y \sin \theta) - \mu t\right) d\mu d\theta$$

Equation (4.9) has been modified slightly. It is not exactly what would be obtained by substituting equation (4.8) into equation (4.1). The part of the spectrum given by (4.8) which involves $\mu + 2\pi/T$ is very small for μ greater than zero. The dominant contribution is given by the other term in this sum of two terms which involve the frequency. In addition, the integral from zero to infinity has been extended to an integral from minus infinity to plus infinity. This essentially uses the term involving $\mu + 2\pi/T$ in (4.8), but it assumes that the spectral components contributed by this term are traveling in the minus x-direction, which appears at first to be in contradiction to the original statement of the problem. However, if the spectrum given by (4.8) is studied, it can be seen that for typical values of D_w of the order of 10 hours and for typical values of the period, say 10 sec, the contributions to the spectrum which are introduced in the wrong direction are negligible and the part neglected in the right direction is also insignificant. Thus all *important* spectral components are traveling in the positive x-direction.

This step then gives equation (4.9). The integral from minus infinity to infinity is easily evaluated, which justifies this step. If the integral of the spectrum given by (4.8) had been written from zero to plus infinity in (4.9), the evaluation of the integral would have been impossible in closed form. By two minor assumptions at this point it becomes possible to obtain a closed solution which will prove to be a remarkably valid solution.

The next step then is to integrate equation (4.9). To this end, the substitutions given by equations (4.10) and (4.11) can be applied

$$\alpha = \mu - \frac{2\pi}{T} \tag{4.10}$$

$$\beta = \frac{\mu^2}{g}\sin\theta \tag{4.11}$$

The inverses of equations (4.10) and (4.11) are

$$\mu = \alpha + \frac{2\pi}{T} \tag{4.12}$$

$$\theta = \sin^{-1}\left(\frac{g\beta}{\left(\alpha + \frac{2\pi}{T}\right)^2}\right) \tag{4.13}$$

The Jacobian of μ and θ can be computed, and the result is that the differential in the new coordinate system is

$$d\mu d\theta = \frac{g}{\sqrt{\left(\alpha + \frac{2\pi}{T}\right)^4 - (g\beta)^2}}\, d\alpha d\beta \tag{4.14}$$

When equations (4.12), (4.13), and (4.14) are substituted into equations (4.9) the result is equation (4.15). This integral can almost be evaluated as it stands except for the term involving the radical in the coefficient of x. The major contributions to this integral occur near values of $\alpha = 0$ and $\beta = 0$. Consequently, the radical can be expanded in a binomial series and higher-order terms can be neglected. The result of this expansion, upon setting the coefficients of higher powers of α and β

equal to zero, is given by (4.16)

$$(4.15)\quad \eta(x,y,t) = \int_{-\infty}^{\infty}\int_{-\infty}^{\infty} \frac{A\sin\left(\alpha\frac{D_w}{2}\right)}{\pi\alpha}\cdot\frac{\sin\left(\frac{\beta W_s}{2}\right)}{\pi\beta}\cdot$$

$$\sin\left(\frac{\left(\alpha+\frac{2\pi}{T}\right)^2}{g}\sqrt{1-\left(\frac{g\beta}{\left(\alpha+\frac{2\pi}{T}\right)^2}\right)^2}\,x+\beta y-\left(\alpha+\frac{2\pi}{T}\right)t\right)d\alpha d\beta$$

$$(4.16)\quad \eta(x,y,t) = \int_{-\infty}^{\infty}\int_{-\infty}^{\infty} \frac{A\sin\left(\alpha\frac{D_w}{2}\right)}{\pi\alpha}\cdot\frac{\sin\left(\frac{\beta W_s}{2}\right)}{\pi\beta}\cdot$$

$$\sin\left[\frac{4\pi^2 x}{gT^2}-\frac{2\pi t}{T}+\frac{\alpha^2 x}{g}+\left(\frac{4\pi x}{gT}-t\right)\alpha-\frac{gT^2 x}{8\pi^2}\beta^2+\beta y\right]d\alpha d\beta$$

The approximation which is used to go from equation (4.15) to equation (4.16) is not the same as the approximation used in stationary phase solutions. The results of the integration will still reduce to the correct initial values at $x = 0$.

The evaluation of the integral given by equations (4.16) is straightforward but tedious from this point onward. A general outline of the steps to be taken will be given in order to aid the reader who wishes to carry them out.

Equation (4.16) can be factored by repeated application of trigonometric identities into a sum of products of integrals and constants (as far as the integrations are concerned). Each term in such a sum is a product of three terms. One term involves x, t, T, and L. The second involves an integral over α and t, T, x, and D_w. The third involves x, y, T, and W_s in an integral over β. Each of the integrals can then be evaluated by standard techniques.

The result is then a function of x, y, and t, and the parameters of the solution, times a sinusoidal simple harmonic wave plus another function of x, y, and t and the parameters of the solution times a cosinusoidal simple harmonic wave. These two functions are essentially amplitudes, and the final solution can then be expressed as a single amplitude (represented by the square root of the sum of the squares of the amplitudes of the sine and cosine terms) times a sinusoidal term with a phase shift which depends on the same coefficients.

Equation (4.17) shows that the final solution can be written as an amplitude which is a product of $F(x,y)^{1/2}$ and $G(x,t)^{1/2}$ times a sinusoidal simple harmonic wave with a phase shift, ψ. This phase shift is a function of the same integrals which result in the terms for the amplitude. It is a slowly varying function which is rather difficult to analyze and which need not be treated in detail.

$$\eta(x,y,t) = \frac{A}{2}\,[F(x,y)]^{1/2}[G(x,t)]^{1/2} \sin\left[\frac{4\pi^2 x}{gT^2} - \frac{2\pi t}{T} + \psi(x,y,t)\right] \tag{4.17}$$

$F(x,y)$ is given by equation (4.18), and $G(x,t)$ is given by equation (4.19). Note, of course, that these functions depend on the parameters of the solutions, namely W_s, D_w, and T (the period of the simple sine wave at the source).

$$F(x,y) = \left[\int_{\sqrt{\frac{4\pi}{gT^2x}}\left(y-\frac{W_s}{2}\right)}^{\sqrt{\frac{4\pi}{gT^2x}}\left(y+\frac{W_s}{2}\right)} \cos\frac{\pi}{2}\,\delta^2 d\delta\right]^2 + \left[\int_{\sqrt{\frac{4\pi}{gT^2x}}\left(y-\frac{W_s}{2}\right)}^{\sqrt{\frac{4\pi}{gT^2x}}\left(y+\frac{W_s}{2}\right)} \sin\frac{\pi}{2}\,\delta^2 d\delta\right]^2 \tag{4.18}$$

$$G(x,t) = \left[\int_{\sqrt{\frac{g}{2\pi x}}\left(\frac{4\pi x}{gT}-t-\frac{D_w}{2}\right)}^{\sqrt{\frac{g}{2\pi x}}\left(\frac{4\pi x}{gT}-t+\frac{D_w}{2}\right)} \cos\frac{\pi}{2}\,\gamma^2 d\gamma\right]^2 + \left[\int_{\sqrt{\frac{g}{2\pi x}}\left(\frac{4\pi x}{gT}-t-\frac{D_w}{2}\right)}^{\sqrt{\frac{g}{2\pi x}}\left(\frac{4\pi x}{gT}-t+\frac{D_w}{2}\right)} \sin\frac{\pi}{2}\,\gamma^2 d\gamma\right]^2 \tag{4.19}$$

The integrals given in equations (4.18) and (4.19) are Fresnel integrals. The upper limits are variable functions of x, T, y, W_s, and D_w, for a particular case, and once the point is chosen at which the integral is to be evaluated, the limits become definite numbers and the value of the integral can be found.

As is stated by equation (4.20), the integral from 0 to ∞ of one of the functions given in (4.18) is equal to ½. Thus the integral from $-\infty$ to $+\infty$ is equal to 1 since the functions are all even functions. In fact the integral from 0 to 8 of either one of the functions is essentially ½, as shown in tabulated values given by Jahnke and Emde [41]

$$\int_0^\infty \sin\frac{\pi}{2}\,\alpha^2 d\alpha = \int_0^\infty \cos\frac{\pi}{2}\,\alpha^2 d\alpha = \frac{1}{2} \tag{4.20}$$

It is easy to show that the solution satisfies the initial values. If y lies between the bounds given by equation (4.6) as x approaches zero in equation (4.18), the result is that the integrals involved must be evaluated from $-\infty$ to $+\infty$. Therefore $F(x,y)$ is equal to 2 if y lies between the appropriate bounds. If y does not lie between the appropriate bounds, the integrals reduce either to the integral from ∞ to ∞ or the integral from $-\infty$ to $-\infty$, and the value is identically 0. Consequently $F(x,y)$ equals 2 as x approaches 0 if y lies between $W_s/2$ and $-W_s/2$, and it equals 0 otherwise.

By exactly the same argument, $G(x,t)$ equals 2 as x approaches 0 if t lies between the bounds given by equation (4.6), and it equals 0 otherwise. Therefore, in equation (4.17) the product of $F(x,y)$ and $G(x,t)$ is equal to 4 and the square root is 2. The amplitude of the wave is therefore A inside the ranges given by (4.6) and it is equal to 0 outside of this range. The phase, ψ, is identically 0 inside the range, and it does not matter what the value is outside the range since the amplitude is 0. Consequently the solution given by equation (4.17) reduces to the initial values given by equation (4.6) as x approaches 0.

The effect of $F(x,y)$ is to keep this disturbance from spreading out in the y-direction as it travels off in the positive x-direction. This function can be evaluated as a function of x and y once and for all for positive x, and the result is a function which is essentially constant in amplitude as y ranges from 0 to $\pm W_s/2$. As y approaches the value $+W_s/2$ or $-W_s/2$, the function begins to oscillate, always remaining positive of course, and, just before reaching the value $W_s/2$, it rises rather rapidly and then falls sharply to 0 beyond the value $W_s/2$. The larger the value of x as y is varied, the more spread out this fluctuation near the border becomes; the wave crests therefore spread out sideways a little as they propagate along in the positive x-direction.

To study this spreading effect, consider equation (4.18) when y has values nearly equal to $W_s/2$. The lower value of the range of integration will be a value near 0, and as y approaches $W_s/2$ from values less than y it will first be negative and then become positive.

The upper limit of integration will be large and positive. It will, in fact, usually be very much greater than 10.

The integral to be studied is thus integrated over a range from a value near 0 to a value of essentially $+\infty$.

Let K be the value of this lower limit of integration in equation (4.18). The square root of F is graphed in Fig. 2 as a function of K. This figure shows that the modulation effect is unimportant for K less than approximately -3 and for K greater than about $+1$. Minor oscillations occur before K equals -1, and the function is down to $2/10$ of full amplitude at $K = 1$ and to essentially 0 at $K = 10$.

Some typical values for the parameters in equation (4.18) can be assumed. Suppose that the wavelength is 200 meters, that y is $11(W_s/2)/10$, and that $W_s/2$ is 200 km. How large must x be in order for the modulation effect to become apparent? By imposing the condition that this lower value of the range of integration be less than 10, the value of x can be solved for, and the result is

$$\frac{x}{2} > \frac{\left(y - \frac{W_s}{2}\right)^2}{100L} \tag{4.21}$$

For the parameters which have been given, the result is that x must be greater than 40 km before the modulation effect can be detected at a point $\frac{1}{10}$ of the half-width of the disturbance to the side of the disturbance.

The modulation effect does not become really important until K is

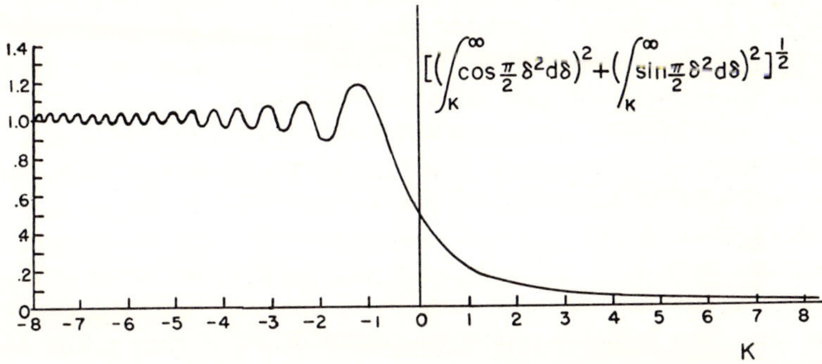

FIG. 2. The upper edge of the side of the envelope. y increases with increasing K. $K = 0$ corresponds to $y = W_s/2$.

less than 1, and when this condition is substituted instead of the condition that K be less than 10, the result is

$$\frac{x}{2} > \frac{\left(y - \frac{W_s}{2}\right)^2}{L} \tag{4.22}$$

For the same values which were given above, this shows that the modulation effect does not produce any important change at a point $\frac{1}{10}$ of the width of the disturbance away from the edge until x exceeds the value of 4000 km. This means, therefore, that the lateral spreading of this disturbance as it travels along in the positive x-direction is such that very little of its effects are observable outside of the lines determined by $y = +W_s/2$ and $y = -W_s/2$ in the x,y-plane.

Consider equation (4.19). It can also be analyzed by the same techniques which were employed to analyze equation (4.18). If the upper limit of integration is set equal to 0 and if the result is solved for t, equation (4.23) is the result. Since $gT/4\pi$ is the group velocity of a wave with a period T, this can also be written as the second equation in (4.23)

$$t = \frac{D_w}{2} + \frac{4\pi x}{gT} = \frac{D_w}{2} + \frac{x}{C_g} \tag{4.23}$$

Similarly, equation (4.24) results when the lower limit of integration is set equal to 0

$$t = -\frac{D_w}{2} + \frac{x}{C_g} \tag{4.24}$$

Thus the modulation envelope as a function of time at a fixed x changes value rapidly at these two values of t. The value given by equation (4.24) is the time required for the forward edge of the disturbance, which started out at $-D_w/2$ sec, to travel to the point x with the group velocity of waves with a period T. Similarly, equation (4.23) represents the rear edge of the disturbance and the time it takes to travel from the origin to the point x, having started at the time $t = +D_w/2$. The modulation as a function of time as observed especially at the line $y = 0$ would therefore have a form exactly like that of Fig. 2, with time increasing toward the left as the forward edge arrives. The reverse occurs as the rear edge arrives at the point of observation.

This solution shows, therefore, that a disturbance of total duration D_w, bounded by the line $y = \pm W_s/2$, starts out from the origin at a certain time. Its forward edge travels with the group velocity of waves with a period T. The disturbance lasts essentially D_w units of time at each point of observation, and its arrival can be predicted. Only a few low waves travel out in advance of the position determined by the group velocity, and only a few others trail to the rear of the back edge of the disturbance as determined by the group velocity. When the disturbance is present, simple harmonic waves of constant period are observed. When it is not present, no waves are observed.

For dimensions of typical storm areas over the oceans, the effect of dispersion is indeed negligible in this case. A discrete spectral component which lasts for 10 or 20 hours and which is 500 or 600 nautical miles wide at the source would essentially be predictable on the basis of full-amplitude sine waves within the bounds described when the disturbance is present, and zero amplitude outside of these bounds except for these minor spreading effects.

The conclusion is, therefore, that a sinusoidal disturbance concentrated at one period, T, will not disperse and will not show any of the so-called period increase of ocean swell over distances comparable to the distance waves travel over the actual ocean surface from an area in which they have been generated. This should not be too surprising because the model certainly does not represent wind generated waves since it is far too regular.

For many hours during the passage of the disturbance, every wave would have exactly the same height as its successors and predecessors and every wave would take exactly the same number of seconds to pass. Wind generated ocean waves are far more irregular than this, and dispersion has a completely different effect when actual wind generated waves are considered.

However, this model is very important in illustrating the effect of group velocity. It shows, for example, that the individual waves form at the rear of the disturbance, grow in amplitude until they reach full amplitude just inside of the rear boundary, determined by the above equations, propagate with their phase speed through the disturbance, go out the front end, and then die down very rapidly and disappear. Thus the individual waves travel through the disturbance, and they are created and finally destroyed by the effect of the modulation envelope.

5. A Stationary Gaussian Process in One Dimension

5.1. Introductory Models

The problem of describing the sea surface in every wave generating area is impossible. The sea surface would be completely different in its detailed structure for each area, although the wind could have the same velocity and the statistical properties of the waves would be the same. The problem of solving a Fourier integral problem along the lines outlined in the last section is also impossible in the practical case because of the extreme irregularity of the waves in the area in which they are generated and because of the impossibility of obtaining the complete set of observations which would be needed. Even given the observations, the time required to carry out the analysis and compute a forecast would be impossibly great. However, as a Fourier integral problem, at least conceivably, the problem could be solved.

The difficulty is that the Fourier integral methods are too cumbersome. Newer techniques are needed which employ the concepts of stationary time series and of statistics. This permits a more general statistical approach which can adequately describe the sea surface in the area in which the waves are generating and which will adequately predict the nature of the swell which propagates out of the area in which the waves are generated.

In order to lead up to these new concepts by a logical series of operations, consider the following model as a very peculiar type of integral. This model is given by

$$I = \int_0^1 f(\mu) \sqrt{d\mu} \tag{5.1}$$

The square root sign over the differential in this equation forces the reader to stop and think of what possible sense this representation can have. Most definitions of this integral would lead to absurd results in that the value would be unbounded. This model is described by Tukey in his appended comments to a paper by St. Denis and Pierson [8], and it is possible to make sense out of this representation.

Suppose that equation (5.1) is defined over a partial sum just as an ordinary Riemann integral is defined over a partial sum. Let this partial sum be obtained by setting $\Delta\mu = 1/N$. The net is then given by $\mu = 0$, $\mu_1 = 1/2N$, $\mu_2 = 1/N$, $\mu_3 = 3/2N$, and so on. The result is given by equation (5.2) where $f(\mu)$ is evaluated at $1/2N$, $3/2N$, and so on up to the point $(2N - 1)/2N$.

$$(5.2)\quad I_N = f\left(\frac{1}{2N}\right)\sqrt{\frac{1}{N} - 0} + f\left(\frac{3}{2N}\right)\sqrt{\frac{2}{N} - \frac{1}{N}} + f\left(\frac{5}{2N}\right)\sqrt{\frac{3}{N} - \frac{2}{N}} + \cdots + f\left(\frac{2N-1}{2N}\right)\sqrt{\frac{N}{N} - \frac{N-1}{N}}$$

The values of $f(\mu)$ are chosen at these points by assigning plus or minus one to the terms according to the results of the toss of a coin. Note that the differential under the square root sign has the value of $\sqrt{1/N}$ for each term which contributes to the sum.

For $N = 2$, $f(\frac{1}{2})$ can be plus or minus one. Also $f(\frac{3}{4})$ can be plus or minus one, and the square root equals $\sqrt{\frac{1}{2}}$. Therefore

I can equal $-\sqrt{2}$ with a probability of 1/4
I can equal 0 with a probability of 1/2
I can equal $+\sqrt{2}$ with a probability of 1/4.

For $N = 4$, $f(\frac{1}{8})$, $f(\frac{3}{8})$, $f(\frac{5}{8})$, and $f(\frac{7}{8})$ are chosen by the same law, and the square root equals $\frac{1}{2}$. Therefore

I can equal -2 with a probability of 1/16
I can equal -1 with a probability of 4/16
I can equal 0 with a probability of 6/16
I can equal $+1$ with a probability of 4/16
I can equal $+2$ with a probability of 6/16.

For $N = 16$, the same laws hold, and the square root equals $\frac{1}{4}$. Therefore

I can equal -4 with a probability of $1/2^{16}$
I can equal -3.5 with a probability of $1/2^{12}$
I can equal -3 with a probability of $15/2^{13}$
I can equal -2.5 with a probability of $35/2^{12}$

I can equal -2 with a probability of $455/2^{14}$
I can equal -1.5 with a probability of $273/2^{12}$
I can equal -1 with a probability of $1001/2^{13}$
I can equal -0.5 with a probability of $715/2^{12}$
I can equal 0 with a probability of $6434/2^{15}$.
From here onward the probabilities descend reading up; the probability of 0.5 is the same as the probability of -0.5, and so on.

For $N = 64$, I can have the values -8, $-7\frac{7}{8}$, $-7\frac{6}{8}$, $-7\frac{5}{8}$, . . . $-\frac{4}{8}$, $-\frac{3}{8}$, $-\frac{2}{8}$, $-\frac{1}{8}$, 0, $\frac{1}{8}$ and so on up to $+8$. The probability that I will equal -8 is given by $1/2^{64}$, and the other probabilities could also be computed if it were so desired.

As N approaches infinity, this expression therefore will take on all the values on the real axis for values of I. The limit is a Gaussian (normal) distribution with a zero mean and a unit variance. Thus it is possible to compute the probability that I will lie between any two numbers on the real axis by integrating the normal distribution with zero mean and unit variance over the distance between these two numbers on the real axis. Of course, the probability that any given value will be obtained is zero in the limit. The distribution becomes essentially continuous.

In this representation, consider what would happen if the square root sign had been omitted from the differential and if $f(\mu)$ were still selected according to the law which was given above. The partial sums would then no longer contain the square root sign as shown in equation (5.2) and, for example, in the last computation where $N = 16$, I would range from -1 to $+1$. In fact, for ever increasing values of N, I would always range from -1 to $+1$, and it could be shown that for N large enough the probability that I would lie between $\pm\epsilon$ could be made to differ from 1 by an arbitrarily small amount. The square root sign over the differential is therefore essential.

The integral defined by equation (5.1) is therefore some number between $-\infty$ and $+\infty$, and the probability that that number will lie within a certain range is determined from the normal probability distribution function. When the ensemble of all possible integrals which can result from all possible combinations of $f(\mu)$ as defined above is considered, then the integral can be thought of as a number with a certain probability that that number will result. Later on, the sea surface will be represented in a somewhat analogous way, and the integral which will be studied will represent all possible sea surfaces with certain basic similar properties. These properties will depend on the spectrum of the sea surface and the properties of a stationary Gaussian process as it will be defined later.

This particular integral as given by equation (5.1) could represent a velocity component of a molecule in a perfect gas, and it is known that these velocity components are normally distributed with a standard deviation which depends upon the temperature. Thus a particular realization of this integral is a particular velocity component, and the ensemble of all possible realizations of the integral describes all possible velocity components of molecules in the gas.

As a second example, which was also given by Tukey in the above reference, consider the integral

$$J(t) = \int_0^1 \cos(t + \epsilon(\mu)) \sqrt{d\mu} \tag{5.3}$$

This integral can be represented by a partial sum over the same net that was used above. The result is

$$\begin{aligned} J_N(t) &= \cos\left(t + \epsilon\left(\frac{1}{2N}\right)\right)\sqrt{\frac{1}{N} - 0} \\ &\quad + \cos\left(t + \epsilon\left(\frac{3}{2N}\right)\right)\sqrt{\frac{2}{N} - \frac{1}{N}} \\ &\quad + \cdots \\ &\quad + \cos\left(t + \epsilon\left(\frac{2N-1}{2N}\right)\right)\sqrt{\frac{N}{N} - \frac{N-1}{N}} \end{aligned} \tag{5.4}$$

Here the quantity $\epsilon(\mu)$ must be assigned in order to make sense out of this integral. Let this quantity be distributed according to the rectangular probability distribution function between 0 and 2π; that is, equation (5.5) should hold where α and $\alpha + d\alpha$ lie between 0 and 2π

$$P(\alpha < \epsilon(\mu) < \alpha + d\alpha) = \frac{d\alpha}{2\pi} \quad \text{for} \quad 0 < \alpha < \alpha + d\alpha < 2\pi \tag{5.5}$$

The partial sum given by equation (5.4) can also be represented as the real part of a sum of vectors in the complex plane

$$J_N(t) = \Re e \sum_{n=1}^{N} \frac{1}{\sqrt{N}} e^{+i\left(t+\epsilon\left(\frac{2n-1}{2N}\right)\right)} \tag{5.6}$$

The result of adding up N terms, each with the same period, namely, 2π, and each with an amplitude of $1/\sqrt{N}$, but with different phases chosen at random, is a sum of vectors which wander around in the plane since each vector points in an arbitrary direction. This is illustrated for $N = 9$, in Fig. 3. The end point of the sum of vectors determines a new vector in the complex plane with a length given by C_N for a partial sum. Since each term in the sum has the same period, all of the vectors will rotate in a counterclockwise direction in the complex plane, and the vector of length C_N will rotate in a counterclockwise direction in the complex plane. The sum will preserve its length, and therefore the result for any particular partial sum and any particular choice of the random phases is a new sinusoidally varying term with an amplitude given by C_N and with a phase given by ϵ_N, both determined by the particular orientation of the vectors which go to make up the total sum. This result is stated in equation (5.7), which also shows that the partial sum can

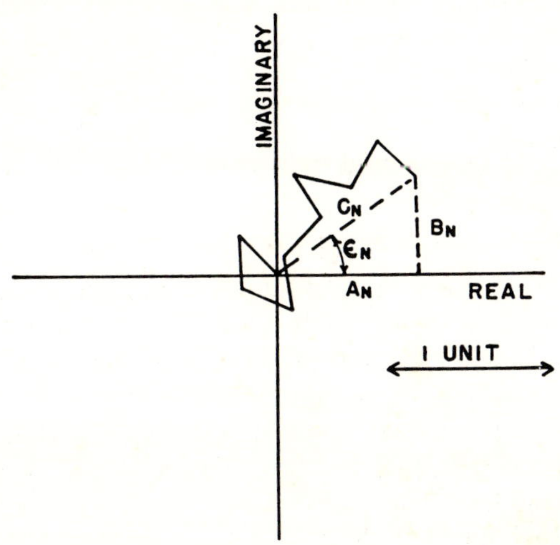

FIG. 3. The complex plane representation for $J_N(t)$ for $N = 9$.

be represented as an amplitude term times a cosinusoidally varying term plus another amplitude term times a sinusoidally varying term

$$J_N(t) = A_N \cos t + B_N \sin t = C_N \cos (t + \epsilon_N) \tag{5.7}$$

These randomly pointed vectors are essentially a two-dimensional random walk problem, and hence the projection of the sum of vectors onto the real axis as N becomes large is distributed with a mean of 0 and a variance equal to ½. This factor, ½, enters by virtue of the way that the probability distribution function of the individual terms is defined. It follows that in the limit as N approaches infinity, the A in equation (5.7) will be distributed according to

$$P(-\infty < A < K) = \frac{1}{\sqrt{\pi}} \int_{-\infty}^{K} e^{-\xi^2} d\xi \tag{5.8}$$

Also the B in equation (5.7) will be distributed according to equation (5.8) since it is independent of A due to the fact that the phase can have any value. The total vector length, C, in the limit is the square root of the sum of the squares of A and B, and standard statistical techniques show, therefore, that C is distributed according to a Chi distribution (or the so-called Rayleigh distribution) as is stated by

$$P(0 < C < K) = \int_0^K 2\xi e^{-\xi^2} d\xi \tag{5.9}$$

A particular realization of $J(t)$ as given by (5.3) is therefore a sinusoidally varying function of time with a period of 2π and with a phase at $t = 0$ arbitrarily distributed somewhere between 0 and 2π. The ensemble of all possible functions can have any phase at $t = 0$, and the probability that the amplitude for a particular realization will be less than K is determined from equation (5.9).

5.2. A Stationary Gaussian Process in One Variable

It now becomes possible to discuss a function which can represent an actual ocean wave record as it would be obtained at a fixed point. The function is thus a function of time. The theory of such a function is also used extensively in communications theory in connection with the study of random noise. It is defined in a similar (but not exactly the same) notation by Rice [33] in his classical paper "The Mathematical Analysis of Random Noise." Other references to this integral can be found in Lévy [32] and in Tukey [34].

In the representation that will be used here, this integral can be given by equation (5.10) where the meaning of the limiting process and the interpretation of the various symbols will be given below.

$$\eta(t) = \int_0^\infty \cos(\mu t + \epsilon(\mu)) \sqrt{[A(\mu)]^2 d\mu} \tag{5.10}$$

There are a number of differences between this expression and equation (5.3). The integral extends from zero to infinity, the frequency of the cosinusoidal term varies as the integral ranges from zero to infinity and there is another term underneath the square root sign which is designated by $[A(\mu)]^2$.

This integral as defined by (5.10) is represented by the limit of a partial sum as given by

$$\eta(t) = \lim_{\substack{\mu_{2r} \to \infty \\ \mu_{2n+2} - \mu_{2n} \to 0}} \sum_{n=0}^{r} \cos(\mu_{2n+1} t + \epsilon(\mu_{2n+1})) \sqrt{[A(\mu_{2n+1})]^2 (\mu_{2n+2} - \mu_{2n})} \tag{5.11}$$

A net of points over the μ-axis is selected. Let them be given by μ_0, μ_1, μ_2, . . . , μ_{2r+2}. Then the value of $[A(\mu)]^2$ at the odd points is selected, it is multiplied by the difference in frequency between the value of μ immediately before and immediately after, and the square root of the result is taken. This determines the amplitude of a sinusoidally varying term. The frequency of the sinusoidally varying term is assigned by the odd numbered value of μ. Finally, a phase chosen at random according to the law given by equation (5.5) is assigned. The result, for a partial sum, is thus a large number of terms each with a different frequency, each with a different phase chosen at random, and each with an amplitude as defined by the value of the square root of the expression in the partial sum.

If the integral from zero to infinity of the spectrum, namely $[A(\mu)]^2$, is equal to a bounded constant, then the representation defined by equations (5.11) and (5.10) has a meaning and it is called a stationary Gaussian process in one variable.

The spectrum which is defined by $[A(\mu)]^2$ is everywhere a positive or zero function defined over the μ-axis from zero to infinity. Its integral over the μ-axis has the value E as given by

$$\int_0^{\infty} [A(\mu)]^2 d\mu = E \tag{5.12}$$

The spectrum has the dimensions of square centimeters $\times$ seconds. Thus when it is multiplied by $d\mu$ the product has the dimensions of square centimeters. The square root then has the dimensions of centimeters. Thus the term under the radical in the definition of the integral assigns an amplitude to each particular term in the sum. As $[A(\mu)]^2$ is varied as a function of frequency, the value of the spectrum is an indication of the amount of potential (or kinetic) energy at that frequency in the wave record. The integral from μ_1 to μ_2 of the spectrum is a measure of the energy between the frequencies μ_1 and μ_2. Hence $[A(\mu)]^2$ will be referred to as an energy spectrum.

Equation (5.11) (at the moment) should not be considered to represent any particular function of time. It should be considered to be an ensemble of an infinite number of different possible functions of time such as can result from the limiting process with all possible choices of the random phase. The statistical properties of this ensemble of functions are studied, and the averages and other statistical quantities which are determined from this equation are found by considering what happens over the set of all the possible phases.

Suppose that the ensemble of functions given by equation (5.10) is looked at at a particular time, t_1. There are many, many different functions and they are all looked at at the particular time, t_1. The values of η at the time t_1 will then have a normal distribution. This is stated by equation (5.13) which states that the ensemble values of $\eta(t_1)$ are normally distributed with a mean of zero and a variance given by $E/2$.

$$(5.13) \qquad P(-\infty < \eta(t_1) < K) = \frac{1}{\sqrt{\pi E}} \int_{-\infty}^{K} e^{-\xi^2/E} d\xi$$

The value of $\eta(t)$ at a time t_j for all possible ensemble realizations of this integral can be studied at the same time that the value of $\eta(t)$ at the time t_k is studied, and the covariance of these two variables can be computed. The covariance as defined by the notation $R(\eta(t_j),\eta(t_k))$ is given by

$$(5.14) \qquad R(\eta(t_j),\eta(t_k)) = \tfrac{1}{2} \int_0^\infty [A(\mu)]^2 \cos (\mu(t_k - t_j))d\mu$$

Thus the values of $\eta(t)$ are correlated when they are observed at different times, and the covariance depends only on the difference in these two times.

In fact, a set of n arbitrary points in time can be selected from equation (5.10). Let them be defined by $t_1, t_2, \ldots, t_j, \ldots, t_k, \ldots, t_n$. These n arbitrary points have a multivariate Gaussian distribution in n variables, such that the variance of each of the points is given by $E/2$ and such that the covariance between any two points, t_j and t_k, is given by equation (5.14).

Moreover, if the set of points $t_1, t_2, \ldots, t_n$ is translated by a linear translation to the set of points $t_1 + t^*, t_2 + t^*, \ldots, t_n + t^*$, the n new points will have the same multivariate Gaussian distribution when studied over the ensemble of all possible functions which can result from equation (5.10).

Therefore the expression given by equation (5.10) represents a stationary Gaussian process; this stationary Gaussian process is such that the multivariate Gaussian distribution for n arbitrary points, no matter how large the value of n, can be exhibited and such that the covariances can be determined. The stationary state is proved by the fact that the probability distribution function of the n variables remains invariant under a translation in time.

A wave record can best be thought of as a particular realization of equation (5.10), and then, by the ergodic theorem, time averages can replace phase or ensemble averages, and the properties of a particular record can be deduced. The study of the properties of a particular realization will be delayed until some additional theoretical foundations have been developed, but the basic concept that a wave record is best represented by equation (5.10) is extremely important.

6. A Stationary Gaussian Process in Three Dimensions

6.1. Formulation

A representation somewhat analogous to the Fourier integral representation given by equation (4.1) can be given which will represent the sea surface as a function of space and time in a form which is also analogous to the equations of the previous section. Consider a free surface of the form

$$(6.1)\quad \eta(x,y,t) = \int_0^\infty \int_{-\pi}^{\pi} \cos\left(\frac{\mu^2}{g}(x\cos\theta + y\sin\theta) - \mu t + \epsilon(\mu,\theta)\right) \cdot \sqrt{[A(\mu,\theta)]^2 d\mu d\theta}$$

This representation must be defined in the same sense that the previous representations were defined, except that now the function is a function of three variables and the spectrum is a function of two variables.

The spectrum is a function of μ and θ, and it has the form $[A(\mu,\theta)]^2$. The spectrum is everywhere greater than or equal to zero. Notice also that it is defined in the entire μ,θ-plane. In an area of wave generation it would probably be zero over one-half of the μ,θ polar coordinate plane if no swell were present. The representation given by equation (6.1), however, also permits the representation of the sea surface when a wave system from a distant storm is running at an angle greater than 180 degrees to the dominant waves in a locally generated wind-driven sea.

The spectrum has the dimensions of cm^2-sec/radian. When it is multiplied by the differential, $d\mu d\theta$, the result has the dimensions of square centimeters, and, when the square root is taken, the result is an amplitude with the dimensions of centimeters. The integral over the spectrum from, say, μ_1 to μ_2, and from θ_1 to θ_2 gives a number with the dimensions of square centimeters. This number then represents that part of the energy (apart from a constant) in the total disturbance associated with the frequencies and directions within the bounds of the range of integration. Hence $[A(\mu,\theta)]^2$ will also be called an energy spectrum. Of course at times for a given choice of μ_1, μ_2, θ_1, and θ_2, the value of the integral will be zero, which means that there is no energy associated with waves traveling in that range of directions with that range of frequencies. As equation (6.2) states, the double integral from 0 to ∞ and from $-\pi$ to π of the spectrum yields the same number E which was defined previously in equation (5.12). This has yet to be proved, but the proof will be given very shortly.

It can be proved that equation (6.1) represents a more general type of stationary Gaussian process in three dimensions. The theorem and the proof are given below (see also [42]).*

$$\int_0^\infty \int_{-\pi}^{\pi} [A(\mu,\theta)]^2 d\theta d\mu = E \tag{6.2}$$

6.2. Statement and Proof of Theorem

THEOREM: *In the limit,* as the mesh of the μ,θ-net in equation (6.3) approaches zero over the full μ,θ-plane, *the n random variables,*

$$\eta(x_i,y_i,t_i) \qquad (i = 1, \ldots, n)$$

which are obtained by considering the ensemble of values of

$$\eta(x,y,t) = \sum_{q=0}^{s} \sum_{m=0}^{r} \cos\left[\frac{\mu^2{}_{2m+1}}{g}(x \cos\theta_{2q+1} + y \sin\theta_{2q+1}) - \mu_{2m+1}t + \epsilon(\mu_{2m+1}, \theta_{2q+1})\right] \cdot \sqrt{[A(\mu_{2m+1},\theta_{2q+1})]^2(\mu_{2m+2} - \mu_{2m})(\theta_{2q+2} - \theta_{2q})} \tag{6.3}$$

at the set of points, x_i,y_i,t_i $(i = 1, \ldots, n)$, scattered arbitrarily in the whole x,y,t-space, *have a multivariate Gaussian distribution* in which the first moments are zero and in which the second moments are given by

$$R(\eta(x_j y_j t_j),\eta(x_k,y_k,t_k)) = \frac{1}{2}\int_{-\pi}^{\pi}\int_0^\infty [A(\mu,\theta)]^2 \cdot \cos\left[\frac{\mu^2}{g}((x_k - x_j)\cos\theta + (y_k - y_j)\sin\theta) - \mu(t_k - t_j)\right] d\mu d\theta \tag{6.4}$$

where k and j range from 1 to n, and where the variances are given when $k = j$ and the covariances are given when $k \neq j$.

The *distribution* of these n random variables *remains unchanged by a translation* of the set of points

$$(x_i y_i t_i) \qquad (i = 1, \ldots n)$$

into the set of points

$$(x_i + x^*,y_i + y^*,t_i + t^*) \qquad (i = 1, \ldots, n)$$

since

$$R(\eta(x_j + x^*,y_j + y^*,t_j + t^*),\eta(x_k + x^*,y_k + y^*,t_k + t^*)) = R(\eta(x_j,y_j,t_j),\eta(x_k,y_k,t_k)) \tag{6.5}$$

Therefore $\eta(x,y,t)$ represents a three-dimensional stationary Gaussian process.

* The results of subsections 6.1 through 6.5 were obtained while conducting research for the Office of Naval Research under contract N onr 285(03). They were first reported to a joint meeting of the American Meteorological Society and the American Statistical Association held in New York City in January, 1954.

PROOF:

To simplify notation, let

$$A_{mq} = \sqrt{[A(\mu_{2m+1},\theta_{2q+1})]^2(\mu_{2m+2} - \mu_{2m})(\theta_{2q+2} - \theta_{2q})} \tag{6.6}$$

and

$$Z^i_{mq} = A_{mq} \cos [g_{mq}(x_i,y_i,t_i) + \epsilon(\mu_{2m+1},\theta_{2q+1})] \tag{6.7}$$

where

$$g_{mq}(x_i,y_i,t_i) = \frac{\mu^2_{2m+1}}{g} (x_i \cos \theta_{2q+1} + y_i \sin \theta_{2q+1}) - \mu_{2m+1}t_i \tag{6.8}$$

and let

$$h^{jk}_{mq} = \frac{\mu^2_{2m+1}}{g} ((x_k - x_j) \cos \theta_{2q+1} + (y_k - y_j) \sin \theta_{2q+1}) - \mu_{2m+1}(t_k - t_j) \tag{6.9}$$

Now consider the double sum represented by equation (6.3). Let the net over $\mu - \theta$ be fixed. Let the set of points x_i, y_i, t_i be fixed. The n random variables under study are then given by

$$\eta(x_i,y_i,t_i) = \sum_{q=0}^{s} \sum_{m=0}^{r} Z^i_{mq} \tag{6.10}$$

after the substitution of equation (6.7) into equation (6.3). The term $\epsilon(\mu_{2m+1},\theta_{2q+1})$ makes Z^i_{mq} a random variable, and everything else in the definition of Z^i_{mq} is a constant. Therefore, for a fixed i, the Z^i_{mq} are independent for different subscripts. Equation (6.10) is therefore a sum of independent random variables for each value of i.

From the properties of trigonometric functions and from the fact that the range of the rectangular distribution of $\epsilon(\mu_{2m+1},\theta_{2q+1})$ is 2π, the probability distribution function of Z^i_{mq} can be shown to be

$$f(Z^i_{mq}) = \frac{1}{\pi} \frac{1}{\sqrt{A^2_{mq} - (Z^i_{mq})^2}} \tag{6.11}$$

for $-A_{mq} < Z^i_{mq} < A_{mq}$, and zero otherwise. The expected value of Z^i_{mq} in (6.11) is zero, and the variance is $A^2_{mq}/2$. Note that it is not necessary to exhibit the joint distribution of the $Z^i_{mq}(i = 1, \ldots, n)$ because of the special procedure which will be used to get the needed moments.

Still for a fixed net, consider the second moment given by

$$R^*(\eta(x_j,y_j,t_j),\eta(x_k,y_k,t_k)) = E^*\left[\left(\sum_{q=0}^{s} \sum_{m=0}^{r} Z^j_{mq}\right)\left(\sum_{q=0}^{s} \sum_{m=0}^{r} Z^k_{mq}\right)\right] \tag{6.12}$$

This can be written as equation (6.13) since $Z^j{}_{mp}$ is independent of $Z^k{}_{mq}$ for different values of the subscripts.* The expectation can be taken under the summation sign since (6.12) is a finite sum.

$$R^*(\eta(x_j,y_j,t_j),\eta(x_k,y_k,t_k)) = \sum_{q=0}^{s} \sum_{m=0}^{r} E^*(Z^j{}_{mq}Z^k{}_{mq}) \tag{6.13}$$

From the simplifying equations given above, $Z^j{}_{mq}$ and $Z^k{}_{mq}$ can be written as

$$Z^j{}_{mq} = A_{mq} \cos [g_{mq}(x_j,y_j,t_j) + \epsilon(\mu_{2m+1},\theta_{2q+1})] \tag{6.14}$$

and

$$Z^k{}_{mq} = A_{mq} \cos [g_{mq}(x_k,y_k,t_k) + \epsilon(\mu_{2m+1},\theta_{2q+1})] \tag{6.15}$$

However, $Z^k{}_{mq}$ can be rewritten as

$$Z^k{}_{mq} = Z^j{}_{mq} \cos (h^{kj}{}_{mq}) - \pm \sqrt{1 - (Z^j{}_{mq})^2} \sin (h^{kj}{}_{mq}) \tag{6.16}$$

where the sign in front of the radical depends on the argument of the cosine term which defines $Z^j{}_{mq}$.

The product $Z^j{}_{mq}Z^k{}_{mq}$ then becomes two terms, one involving $(Z^j{}_{mq})^2$ times $\cos (h^{kj}{}_{mq})$ and the other involving $Z^j{}_{mq}(\pm \sqrt{1 - (Z^j{}_{mq})^2})$ times $\sin (k^{kj}{}_{mq})$. The expected value of the first term is $(A^2{}_{mq}/2) \cos (h^{kj}{}_{mq})$, and the expected value of the second term is zero since its value is simply the expected value of $A^2{}_{mq} \sin \alpha \cos \alpha$, where α is rectangularly distributed over a range from a constant to a constant plus 2π.

Equation (6.13) therefore becomes

$$R^*(\eta(x_j,y_j,t_j),\eta(x_k,y_k,t_k)) = \sum_{q=0}^{s} \sum_{m=0}^{r} \frac{A^2{}_{mq}}{2} \cos (h^{kj}{}_{mq}) \tag{6.17}$$

A return to the original definition of the various terms involved then yields

$$\begin{aligned} (6.18)\quad & R^*(\eta(x_j,y_j,t_j),\eta(x_k,y_k,t_k)) \\ &= \sum_{q=0}^{s} \sum_{m=0}^{r} [A(\mu_{2m+1},\theta_{2q+1})]^2 \cdot \\ &\cos \left[\frac{\mu^2{}_{2m+1}}{g} ((x_k - x_j) \cos \theta_{2q+1} + (y_k - y_j) \sin \theta_{2q+1}) - \mu_{2m+1}(t_k - t_j) \right] \\ &\cdot [\mu_{2m+2} - \mu_{2m}][\theta_{2q+2} - \theta_{2q}] \end{aligned}$$

* The symbol E^* stands for "the expected value of." It is not the same as the E of the other parts of this paper.

Considered all by itself, the following limiting process can certainly be applied to equation (6.18), and the result as stated by equation (6.19) is the ordinary Riemann integral defined by equation (6.4).

$$\lim_{\substack{s\to\infty\\ r\to\infty\\ \mu_{2m+2}-\mu_{2m}\to 0\\ \theta_{2q+2}-\theta_{2q}\to 0\\ \mu_{2r}\to\infty}} R^*(\eta(x_j,y_j,t_j),\eta(x_k,y_k,t_k)) = R(\eta(x_j,y_j,t_j),\eta(x_k,y_k,t_k)) \tag{6.19}$$

It is now possible to apply Theorem 21a in Chapter 10 of Cramer [43]. The theorem applies to the sum of n dimensional random variables each of which has zero means and finite second moments, provided that the generalized Lindeberg conditions hold. A sum of N such variables, divided by $\sqrt{N}$, as N approaches infinity, is then shown to be multivariate Gaussian with a zero mean and with second moments equal to the sums of the second moments of the original random variables divided by N.

The mesh of the net in the results of this paper plays the same role as the N in the above theorem. Let $N = (s + 1)(r + 1)$. Then, by stretching the range of definition of equation (6.11) by the amount $\sqrt{N}$, the new range will remain finite as the number of terms increases, and the sum divided by $\sqrt{N}$ will be the same as obtained before in (6.10). The moments will be N times bigger, but they will be divided by N and hence the same result will be obtained. Even the expanded distributions behave properly for large ranges of the variable since they are identically zero outside of a certain finite range and therefore the generalized Lindeberg condition is satisfied.

Cramer's theorem [43] therefore applies directly to the n random variables given in equation (6.10). Therefore, as N approaches infinity, which corresponds to shrinking the mesh of the net in the μ,θ-plane uniformly over the whole plane, the n values of $\eta(x_i,y_i,t_i)$, $(i = 1, \ldots, n)$, have a multivariate Gaussian distribution with zero means and with second moments given by equation (6.4).

Also the substitution of $x_j + x^*$, $y_j + y^*$, $t_j + t^*$ and $x_k + x^*$, $y_k + y^*$, $t_k + t^*$ for x_j, y_j, t_j and x_k, y_k, t_k into equation (6.4) yields the same right hand side.

Therefore, the process is a three-dimensional stationary Gaussian process.

6.3. Reduction to the One-Variable Case as a Function of Time

Let the set of points in the theorem given above be defined so that $x_1 = x_2 = x_3 = \cdots = x_n$ and $y_1 = y_2 = y_3 = \cdots = y_n$. Under these conditions, equation (6.4) reduces to a form similar to equation (5.14), except for the fact that an integral over θ is still involved. However, if the spectrum as a function of frequency is defined by the integral from $-\pi$ to π of the spectrum as a function of frequency and direction, as is stated by

$$\int_{-\pi}^{\pi} [A(\mu,\theta)]^2 d\theta = [A(\mu)]^2 \tag{6.20}$$

then equation (6.4) and equation (5.14) become identical for this special case. This also shows that the E defined by equation (6.2) is equal to the E defined by equation (5.12).

The free surface as a function of x, y, and t under these conditions reduces to a free surface as a function of time alone, as given by equation (5.10). Thus the ensemble of representations of the free surface as a function x, y, and t as given by equation (6.1) reduces to the ensemble of representations of the free surface as a function of time alone when the set of points under study is such that the points are located at a fixed point in the x,y-plane. The spectrum as a function of frequency and direction reduces to a spectrum as a function of frequency, which is to be expected because the observation of any type of wave at one point as a function of time tells the observer nothing about the direction toward which the wave is traveling. In fact, from the information available, the motion could be in part a standing wave. The proof of the statements made in section 5 about the stationary Gaussian process as a function of one variable could be given by following exactly the same proof which was used above in the statement and proof of the theorem for the more general case of the stationary process in three dimensions. However, the more general case was proved, and then the special case is obtained immediately from the above considerations.

6.4. Reduction to the One-Variable Case Along a Line in Space at an Instant of Time

Suppose that the ensemble of possible wave states given by equation (6.1) is observed at a set of points x,y,t, such that t has the same value for each point. Also suppose that the pairs of points x_1,y_1, x_2,y_2, x_3,y_3, $\ldots$, x_n,y_n lie along a line in the x,y-plane. It can be shown that the ensemble properties of these points can also be determined and that the result is a spectrum which is again a function of one variable.

The proof of this statement is as follows:

Let the set of points in the x,y,t-space be points which lie on the line determined from the intersection of the planes, $t = t_1$ and

$$y \cos \theta_D - x \sin \theta_D = y_1'$$

The n random variables then become

$$\eta(x_i,y_i,t_i) \qquad (i = 1, \ldots, n)$$

where the x_i are a linear function of the y_i.

Consider the new coordinate system given by

$$\begin{aligned} x' &= x \cos \theta_D + y \sin \theta_D \\ y' &= -x \sin \theta_D + y \cos \theta_D \end{aligned} \tag{6.21}$$

which is a counterclockwise rotation of the x,y-coordinate axes through θ_D radians. The inverse is given by

$$\begin{aligned} x &= x' \cos \theta_D - y' \sin \theta_D \\ y &= x' \sin \theta_D + y' \cos \theta_D \end{aligned} \tag{6.22}$$

When equation (6.21) is substituted into equation (6.4), the result is equation (6.23) since $t_k - t_j = 0$ and $y_k' - y_j' = 0$ for all k and j

$$\begin{aligned} (6.23) \quad & R(\eta(x_j',y_1',t_1),\eta(x_k',y_1',t_1)) \\ & = \frac{1}{2}\int_0^\infty \int_{-\pi}^{\pi} [A(\mu,\theta)]^2 \cos \left[\left(\frac{\mu^2}{g} \cos(\theta - \theta_D) \right) (x_k' - x_j') \right] d\mu d\theta \end{aligned}$$

Now let

$$\nu_0 = \frac{\mu^2 |\cos(\theta - \theta_D)|}{g} \tag{6.24}$$

and

$$\theta_0 = \theta - \theta_D \tag{6.25}$$

The inverses are

$$\mu = \left(\frac{g\nu_0}{|\cos \theta_o|} \right)^{1/2} \tag{6.26}$$

and

$$\theta = \theta_0 + \theta_D \tag{6.27}$$

The Jacobian of (6.26) and (6.27) is given by

$$\frac{\partial(\mu,\theta)}{\partial \nu_0 \partial \theta_0} = \frac{1}{2} \left(\frac{g}{\nu_0 |\cos \theta_0|} \right)^{1/2} \tag{6.28}$$

When the transformation of variables given by (6.24) and (6.25) is applied to equation (6.23), the result is

$$\begin{aligned} (6.29) \quad & R(\eta(x_j',y_1',t_1),\eta(x_k',y_1',t_1)) \\ & = \int_0^\infty \int_{-\pi}^{\pi} \left[A\left(\left(\frac{g\nu_0}{|\cos \theta_0|} \right)^{1/2}, \theta_0 + \theta_D \right) \right]^2 \cos[\nu_0(x_k' - x_j')] \frac{\partial(\mu,\theta)}{\partial \nu_0 \partial \theta_0} d\theta_0 d\nu_0 \\ & = \frac{1}{2} \int_0^\infty [A^*(\nu_0;\theta_D)]^2 \cos[\nu_0(x_k' - x_j')] d\nu_0 \end{aligned}$$

In equation (6.29), the integration over θ_0 can be carried out and the result defined in terms of ν_0 with a parameter θ_D, as is done in the second expression in (6.29). $[A^*(\nu_0,\theta_D)]^2$ is defined by

$$(6.30) \quad [A^*(\nu_0;\theta_D)]^2 = \int_{-\pi}^{\pi} \left[A\left(\left(\frac{g\nu_0}{|\cos\theta_0|} \right)^{1/2}, \theta_0 + \theta_D \right) \right]^2 \left(\frac{g}{\nu_0 |\cos\theta_0|} \right)^{1/2} d\theta_0$$

The spectrum, $[A^*(\nu_0;\theta_D)]^2$, is a function of frequency alone in terms of variations as a function of x' along the line $y' = y_1'$ at the time $t = t_1$. A representation analogous to equation (6.29) as a function of x' is therefore possible and is given by

$$(6.31) \quad \eta(x',y_1',t_1) = \int_0^\infty \cos\left[\nu_0 x' + \epsilon(\nu_0)\right] \sqrt{[A^*(\nu_0;\theta_D)]^2 d\nu_0}$$

The results of what has just been given above can be explained by considering that along a line in the x,y-plane the various infinitely long crested waves which go to make up a partial sum which represents equation (6.1) need not be such that the crest is perpendicular to this line. Therefore, different wave frequencies as a function of space are obtained, depending upon the angles that the crests make with the line in the x,y-plane. Different time frequencies map into different space frequencies because of the effect of the θ variable, and the same time frequencies map into different space frequencies because of the effect of the θ variable. The transformation given by equations (6.24) and (6.25) assigns the correct space frequencies for each time frequency and direction as measured with respect to θ_D, which is the angle which this line in the x,y-plane makes with the x-axis. The final result is thus a spectrum with frequencies which are related to a function of distance along a line, and along this line the process is a stationary Gaussian process with respect to distance.

6.5. Reduction to One Variable as Observed at a Moving Point

A special case applicable to ship motions can also be studied where the set of points x_1,y_1,t_1; x_2,y_2,t_2; . . . ; x_n,y_n,t_n is such that the points lie on a line in the x,y,t-space. This line is a line determined by a point moving along a straight line in the x,y-plane with a velocity v. The transformations which must be employed to study this case are given by St. Denis and Pierson [8], and they will not be repeated here.

In connection with geophysical problems, there are a number of wave recorders designed to record the waves as a ship encounters them. The effect of the motion of the ship is partially removed, but if the ship is under way the frequencies which are obtained from an analysis of this record have very little relationship to the true frequencies of a wave record as observed as a function of time at a fixed point, and returning to a spectrum which will describe a wave record as a function of time at a fixed point is extremely difficult. For such records to be of value, the ship should heave to and remain stationary as the waves pass the ship.

6.6. Comparison with Fourier Integrals

In equation (4.1) the free surface was represented as a Fourier integral in order to solve initial value problems. This integral looks considerably like equation (6.1). However, there are a great many differences which must be emphasized. The free surface as a function of space and time, when represented by a Fourier integral, must approach zero at the boundaries of the space x,y,t; that is, as x, y, or t approaches infinity, the value of $\eta(x,y,t)$ must go to zero. In addition, the integral is evaluated over spectra which are in phase, that is, the integral is a sum of many sinusoidal terms which are all in phase at the origin. The spectrum is a bounded spectrum, and there is no square root sign involved over the differential. Thus, with the ability to be in phase near $x = y = t = 0$, the sum of terms represented by the integral can add up to a finite disturbance near the origin. However, at great distances from the origin, the various terms in the sum get out of phase and the representation dies down to zero.

Equation (6.1) represents an ensemble of many different sea surfaces. If every term were in phase (which they are not because of the random phase given by $\epsilon(\mu,\theta)$), the result would be that the free surface would be infinite at a point, say the origin, as the mesh of the net went to zero. However, the random phases (and the fact that one considers the ensemble of all possible values) make it possible to show that the value of free surface at the origin is a number from a normal distribution with a zero mean and a variance given by $E/2$. Also, the representation given by equation (6.1) never dies down to zero for large values of x and y, and t. This statement must follow by virtue of the fact that it has been shown that this representation is stationary and that a translation of a given set of points to any part of the x,y,t-space still has the same multivariate Gaussian distribution. The square root sign over the differential combined with the effect of the random phases results in a free surface as represented by equation (6.1) which never dies down to zero over any large part of the x,y,t-space. This means that a particular realization of a representation such as equation (6.1) could theoretically go on forever as a function of time at a fixed point and that it would be defined over the entire x,y-plane; and hence, in a practical geophysical application, this would imply, possibly, that the representation would be defined to be the same over the whole North Atlantic Ocean.

The interpretation to be given to equations (4.1) and (6.1) is thus a question of scale. If one is considering the state of the sea, say, in the analysis of conditions over the entire North Atlantic Ocean, then the free surface should be considered to be related to equation (4.1). The disturbance created by a given area of high winds is essentially zero outside the area of high winds, if the waves have not started to travel out of the area, and there should be parts of the ocean where the sea is relatively calm. Thus for gross-scale effects it must be considered that the free surface is defined in some way such that it does not have the same statistical properties over an area the size, say, of the North Atlantic Ocean.

On the other hand, consider an area of the sea surface 500 miles wide and 400 or 500 miles long, over which the wind has been blowing with a velocity of 20 knots for two or three days. Then, within this area, the free surface can be thought of as one of the possible realizations of equation (6.1), with the reservation that if the observer moves out of this area, or observes the waves at some other time, the representation is no longer valid.

Now, certainly it would not be possible to observe the free surface in an area as large as the one described with sufficient accuracy to formulate a Fourier integral. However, it is possible to observe a portion of the area described with enough accuracy to obtain a reliable estimate of the energy spectrum of equation (6.1). The statistical validity of the estimate of the energy spectrum given in equation (6.1) can be determined, and it is not too difficult to obtain enough data to establish some of the properties of a sea surface defined by equation (6.1) in a storm area.

Of course, not every bump and hollow or every short crested wave is observed. This procedure replaces the particular sea surface which is being observed by an ensemble of an infinite number of other possible sea surfaces with exactly the same energy spectrum and with exactly the same statistical properties for a given steady state.

7. The Ergodic Theorem

7.1. Definition and Application

The statistical properties of the ensemble of all possible sea surfaces with a given energy spectrum have been studied above. For a stationary Gaussian process, the ergodic theorem states that time and space analyses yield the same statistical properties as does an ensemble analysis. Consequently, wave records, obtained as a function of time at a fixed point, stereo-aerial photographs of the sea surface, and ship motion records are all examples of particular realizations of (quasi-) stationary Gaussian processes. The developments in communication theory of Rice [33] and Tukey [34] are thus applicable to the analysis of ocean wave records.

With the use of the partial sum given in equation (5.11) to approximate a particular realization, before the limit is approached, it is easily proved for a set of phases assigned at random that the following equation is valid where $p = t_k - t_j$. The result is the same as (5.14).

$$\lim_{\bar{T}\to\infty} \frac{1}{\bar{T}} \int_{t^*}^{t^*+\bar{T}} \eta(t)\eta(t + (t_k - t_j))dt = \frac{1}{2}\int_0^\infty [A(\mu)]^2 \cos(\mu(t_k - t_j))d\mu = \frac{Q(p)}{2} \tag{7.1}$$

Since

$$[A(\mu)]^2 = \frac{2}{\pi}\int_0^\infty Q(p) \cos \mu p \, dp \tag{7.2}$$

the energy spectrum of a particular wave system can be found by forming lagged products of a given wave record with itself and evaluating the even Fourier cosine transform of the result.

Similar methods of analysis make it possible to find $[A^*(\nu_0;\theta_D)]^2$ for a particular state of the sea, and the spectrum of the motion of a particular ship in a particular seaway. Stereo-aerial photographs should make it possible to determine the full spectrum $[A(\mu,\theta)]^2$ for a particular seaway, as shown by Marks [44].

The actual computations must be carried out carefully, following the procedures given by Tukey [34] as applied by Pierson and Marks [45] to wave records. Additional records have been analyzed since then, and they all substantiate the theoretical spectra derived by Neumann [38], which will be discussed later.

7.2. Potential and Kinetic Energy

The potential energy averaged over time for a unit sea surface area is given by

$$\overline{\text{P.E.}}^t = \lim_{\bar{T}\to\infty} \frac{\rho g}{2} \int_{t^*}^{t^*+\bar{T}} (\eta(t))^2 dt = \frac{\rho g}{4} \cdot E \tag{7.3}$$

This shows that, apart from a constant, E is related to the potential energy of a given state of the sea and that the fraction of the total energy within a given frequency range equals the integral over that frequency range of the energy spectrum. This is why the spectrum is called the energy spectrum. Also, since the kinetic energy (integrated over depth) equals the potential energy, and from the properties derived in section

6, the results stated in equation (7.4) can also be proved

$$\overline{\text{P.E.}}^t = \overline{\text{K.E.}}^t = \overline{\text{P.E.}}^{x'} = \overline{\text{K.E.}}^{x'} \tag{7.4}$$

In which x' is any line on the sea surface at an instant of time. Equation (7.4) is true only for a short crested Gaussian sea surface. For more elementary short crested waves, it definitely is not valid.

There is no difficulty in extending these results to other properties of the waves. The velocity components at any depth and the pressure at any depth are easily derived.

8. A Definition of a Wind Generated Wave and of a Wind Generated Gravity Wave System

Ocean waves and ocean wave records have been studied for many years but the studies lacked the unifying concept of what ocean waves actually were. All available evidence points to the hypothesis that ocean waves can in many cases be represented by the stationary Gaussian process in three variables described in section 6.

The record of the waves as they pass a fixed point as a function of time alone was first shown to be Gaussian by Rudnick [5]. Since then many observations have verified this property. For high waves, especially in shallow water, the crests are higher and more peaked and the troughs are shallower and flattened. Thus such waves are not strictly Gaussian due to the effect of nonlinearity on the higher frequencies, as reported by Birkhoff and Kotik [6]. However, for actual waves the results of the application of the Gaussian hypothesis to such waves yields more realistic results than any known nonlinear (and hence non-Gaussian) solution.

An attempt will now be made to answer the question posed by Mason in 1949. The question is "What is an ocean surface wave?"

The answer is: "A wind generated ocean surface wave is a bump on the water."* This answer is all that really can be given. Knowledge of just one wave is not enough to permit any knowledge of the properties of a wave system.

A wind generated ocean surface wave never occurs alone. There are always many waves present, and these waves form an ocean wave system. The question posed by Mason must therefore be restated as: "What is an ocean wave system?"

The answer to this question is: Over an area of the ocean, perhaps 500 miles on a side, and often for several days, it appears that a wind generated wave system can be represented (apart from minor nonlinear effects) as a particular realization of a quasi-stationary Gaussian process in three variables (if the water for practical purposes is infinitely deep)

* The definition of an ocean wave was given to the author by J. V. Hall. Many others have undoubtedly used it.

which is completely characterized in a statistical sense by a spectrum of the form $[A(\mu,\theta)]^2$.

For shallow water, variable winds, and problems in propagation, the concept of a quasi-stationary Gaussian process needs to be defined more precisely, but it is possible to generalize to a complete forecasting theory by treating the energy spectrum as a slowly varying function of space and time. The results which are given in sections 9 through 15 are therefore based on the assumption that the wind generated wave system is a particular realization of a quasi-stationary Gaussian process.

9. Practical Wave Statistics

9.1. The Results of Rice

Rice [33] has derived and presented many of the statistical properties of Gaussian noise. All of his results can be applied to ocean wave records. Pierson [46, 47] has applied the results of Rice to a study of ocean wave records and the properties of ocean waves.

The envelope of a Gaussian wave record has a probability distribution function given by equation (9.1) if the spectrum is narrow. Even for irregular sea waves with a broad spectrum, the amplitudes of the waves (crest to mean sea level, and trough to mean sea level) are approximately distributed according to this probability distribution function

$$P(0 < a < K) = \int_0^K \frac{2\xi}{E} e^{-\xi^2/E} d\xi \tag{9.1}$$

The crest-to-trough wave height distribution is very closely given by doubling the values of the probability distribution function of the amplitudes.

The derivation of the probability distribution function of the wave heights was given independently by Barber [48], and it was also presented by Longuet-Higgins [49]. Longuet-Higgins extended the analysis to discover the probability distribution function of the highest wave out of N waves.* The properties of this probability distribution function have been verified conclusively on pressure wave records by Watters [51]. Putz [52, 53] recently extended certain aspects of this theory, for cases where the wave record is extremely irregular, to obtain the probability distribution of the crest-to-trough heights.

A complete explanation of the ratios found by Seiwell [54] and Weigel [55] can be given by the use of equation (9.1). The average wave height equals 1.77 $\sqrt{E}$. The significant height equals 2.83 $\sqrt{E}$, and the average height of the one-tenth highest waves equals 3.60 $\sqrt{E}$. Equation

* See also Pierson, Neumann, and James [50].

(9.1) is truncated at the appropriate point, and the first moment about the origin of the truncated distribution is computed to obtain these values.

The significant height of the waves can be estimated visually or computed easily from a wave record. This permits an estimate of E but, since spectra of many different shapes can have the same area under them, this one number is not enough to completely characterize the waves.

The average time interval between successive waves (where a wave is defined to be a complete cycle from a zero up cross through a crest, down through a zero down cross to a trough and finally back to a zero up cross) observed at a fixed point is given by

$$\bar{\bar{T}} = 2\pi \left[\frac{\int_0^\infty [A(\mu)]^2 d\mu}{\int_0^\infty \mu^2 [A(\mu)]^2 d\mu} \right]^{1/2} \tag{9.2}$$

The average time interval between successive maxima is given by equation (9.3). A maximum is any part of a wave record with a horizontal tangent and with positive curvature. A maximum can thus occur in a part of the trough of a larger wave.

$$\bar{\bar{T}}_{\max} = 2\pi \left[\frac{\int_0^\infty \mu^2 [A(\mu)]^2 d\mu}{\int_0^\infty \mu^4 [A(\mu)]^2 d\mu} \right]^{1/2} \tag{9.3}$$

The values of $\bar{\bar{T}}$ and of $\bar{\bar{T}}_{\max}$ are in a sense average "periods" of the waves. Again, many spectra can produce the same values of $\bar{\bar{T}}$, and hence this number cannot adequately describe a wave record.

Similar equations can also be applied to the sea surface to obtain the average distances between successive waves and the average distance between successive maxima along a line at an instant of time. For a spectrum which is symmetrical about $\theta = 0$, and for waves observed on a line $y =$ constant, the values are given by

$$\bar{\bar{L}} = 2\pi \left[\frac{\int_0^\infty [A^*(\nu_0;0)]^2 d\nu_0}{\int_0^\infty \nu_0^2 [A^*(\nu_0;0)]^2 d\nu_0} \right]^{1/2} \tag{9.4}$$

$$\bar{\bar{L}}_{\max} = 2\pi \left[\frac{\int_0^\infty \nu_0^2 [A^*(\nu_0;0)]^2 d\nu_0}{\int_0^\infty \nu_0^4 [A^*(\nu_0;0)]^2 d\nu_0} \right]^{1/2} \tag{9.5}$$

There are many additional important results of Rice [33] which can be applied to the analysis of wave records. A review of this reference by all concerned with ocean wave theory is strongly recommended.

9.2. *Results of Wooding* and Barber†*

Wooding and Barber have studied the probability distribution function of the amplitudes of the horizontal tangents to a wave record. The results are more general than the ones given in equation (9.1). For more advanced problems in which such methods would be needed, this work provides much important information.

Equation (5.10) was studied by representing it as the real part of a revolving vector in the complex plane. This complex vector accelerates, decelerates, grows, and shrinks in the complex plane. It can even back up for a short time while tracing out a loop. Its vector length is distributed according to (9.1). Thus, if the vector is not behaving too erratically, the maximum value of its projection on the real axis is a maximum when the vector passes through the real axis, and the probability distribution of the horizontal tangents of $\eta(t)$ coincides with the probability distribution of the envelope given above.

The velocity vector can be defined which represents the velocity of the tip of the complex vector referred to above. A horizontal tangent to $\eta(t)$ occurs when the real part of this velocity vector is zero. The probability distribution of the values of $\eta(t)$ at the times when this condition is satisfied is then the distribution of the horizontal tangents to $\eta(t)$.

According to Wooding and Barber, the probability, dp, that a horizontal tangent (either at a minimum, a maximum, or a point of inflection) will occur between the values of x and $x + dx$ is given by equation (9.6), which is an even function of x,

$$dP = \left[a_1 e^{-a_2 x^2} + a_3 \left[\int_{-a_4 x}^{a_4 x} e^{-\xi^2}\, d\xi\right] |x| e^{-x^2/E}\right] dx \tag{9.6}$$

The values of the constants in equation (9.6) are given by

$$a_1 = \sqrt{\frac{E|E_2| - |E_1|^2}{\pi E^2 |E_2|}} \tag{9.7}$$

$$a_2 = \frac{|E_2|}{E|E_2| - |E_1|^2} \tag{9.8}$$

$$a_3 = \frac{|E_1|}{E\sqrt{E|E_2|\pi}} \tag{9.9}$$

$$a_4 = \frac{|E_1|}{\sqrt{|E_2|E^2 - E|E_1{}^2|}} \tag{9.10}$$

The symbol E is defined elsewhere, and the symbols E_1 and E_2 are

* Robin A. Wooding (Underwater Research Laboratory, D.S.I.R., P.O. Box 22, Devonport, Auckland, New Zealand).

† Norman F. Barber (Oceanographic Observatory, Kelburn, Wellington, New Zealand), unpublished results from a personal communication dated March 16, 1953.

defined by evaluating $E_1(\tau)$ and $E_2(\tau)$ as given by equations (9.11) and (9.12) and setting $\tau = 0$

$$E_1(\tau) = i \int_0^\infty e^{i\mu\tau}\mu[A(\mu)]^2 d\mu \tag{9.11}$$

$$E_2(\tau) = - \int_0^\infty e^{i\mu\tau}\mu^2[A(\mu)]^2 d\mu \tag{9.12}$$

For a spectrum which covers only a narrow band of frequencies, and which is zero outside of this band, the quantity $E|E_2| - |E_1|^2$ is essentially zero, and (9.6) reduces to $(|x|/E) \exp(-x^2/E)$, which is the distribution given by (9.1) halved in amplitude and reflected in the origin to produce a wing-shaped even function.

If another portion at a very high frequency is added to the above narrow-band spectrum, the constants a_1 and a_2 can dominate because E_2 is large and the distribution reduces to a normal distribution with a variance of $E/2$ since the ripple effectively samples the swell at very brief time intervals.

For problems in microwave propagation and other such phenomenon, these results will prove valuable. For large-scale effects, an observer, or the present type of pressure wave recorder, would ignore these features and the probability distribution would be like that given in (9.1).

10. Wave Generation and Theoretical Energy Spectra

10.1. Importance

Sometimes the theoretical developments in a branch of science outrun the observational data on which the science is based. Then it is necessary to concentrate on the observational data in order to catch up with the theoretical studies which have been made. This is the present status of the theory of wind generated gravity waves.

The number of wave records which have been adequately analyzed to obtain the energy spectrum, $[A(\mu)]^2$, is very small. Thus the variation of the spectrum as a function of wind velocity and duration or fetch is not known from the direct computation of many spectra. This lack of data will be remedied in the near future by the data from an electronic wave spectrum analyzer recently developed for the Beach Erosion Board and from an electronic analyzer which has been built for the Division of Oceanography of the Hydrographic Office.

Neumann [36–39] has taken the many visual observations which he made on the *M.S. Heidberg* and combined them with other data to derive the energy spectrum of the waves as a function of wind velocity and either the duration or the fetch. The number of ways in which this theoretical spectrum agrees with the observed properties of wind generated gravity waves is continuously increasing, and no major discrepancy between theory and observation has yet been discovered.

10.2. The Derivation

If a wave record taken in deep water (compared to the important periods present) is studied carefully, it is seen that there is a wide range of variation in the time intervals between individual crests, $\tilde{T}$. If a wave with a low value of the "period" $\tilde{T}_1$ occurs, then its height is produced by the phase reinforcement of many low sinusoidal waves with spectral periods in the range from $\tilde{T}_1 - (\Delta T)/2$ to $\tilde{T}_1 + (\Delta T)/2$. If a wave with a high value of the "period" $\tilde{T}_2$ occurs, then its height is produced by the phase reinforcement of many low sinusoidal waves with spectral periods in the range from $\tilde{T}_2 - (\Delta T)/2$ to $\tilde{T}_2 + (\Delta T)/2$. Of course ΔT need not be a constant; nevertheless, the height associated with a given apparent "period" reflects the nature of the amount of energy in the spectrum within the range of spectral periods near the value $\tilde{T}$.

From a number of such wave records where the sea was well developed, Neumann plotted the log of $H^*/\tilde{T}^2$ versus the ratio $(\tilde{T}/v)^2$, where v was the velocity (in meters/second) of the wind, measured at anemometer level, which generated the waves. His results are shown in Fig. 4. A definite upper bound exists for the $H^*/\tilde{T}^2$ value for each value of $(\tilde{T}/v)^2$. The line forming the upper bound of these values is given by

$$\frac{H^*}{\tilde{T}^2} = K_1 e^{-2.438(\tilde{T}/v)^2} \tag{10.1}$$

This relationship is also substantiated by values determined from bands of swell and in young seas where the spectrum of the waves is only a part of the spectrum of a fully developed sea. The empirical factor 2.438, determined from Fig. 4, is equal to $(g/2\pi)^2$ with g in meters per second per second and with v in meters per second. Consequently, the characteristic height of the waves in a narrow band of frequencies is

$$H^* = K_1 \tilde{T}^2 e^{-(g\tilde{T}/2\pi v)^2} \tag{10.2}$$

Equation (10.2) states that the height of a wave with a low "period"* is essentially dependent on the square of that "period" (or some sort of "wavelength") if that "period" is low in comparison with a "period" whose phase velocity equals the wind velocity. The characteristic height thus first increases as the square of the "period" and then rapidly

* In what follows, the word "period" ($\tilde{T}$) in quotes will designate the nonmathematical meaning of the word. The period of a simple harmonic progressive wave (T) is quite a different thing, and a spectral period is also something different.

decreases as $g\tilde{T}/2\pi v$ approaches unity because the wind cannot supply energy to these waves as easily.

Also, the square of the characteristic wave height which, apart from a proportionality constant, equals the potential energy averaged over time for the same period band is

$$(H^*)^2 = K_1{}^2\tilde{T}^4e^{-2(g\tilde{T}/2\pi v)^2} \tag{10.3}$$

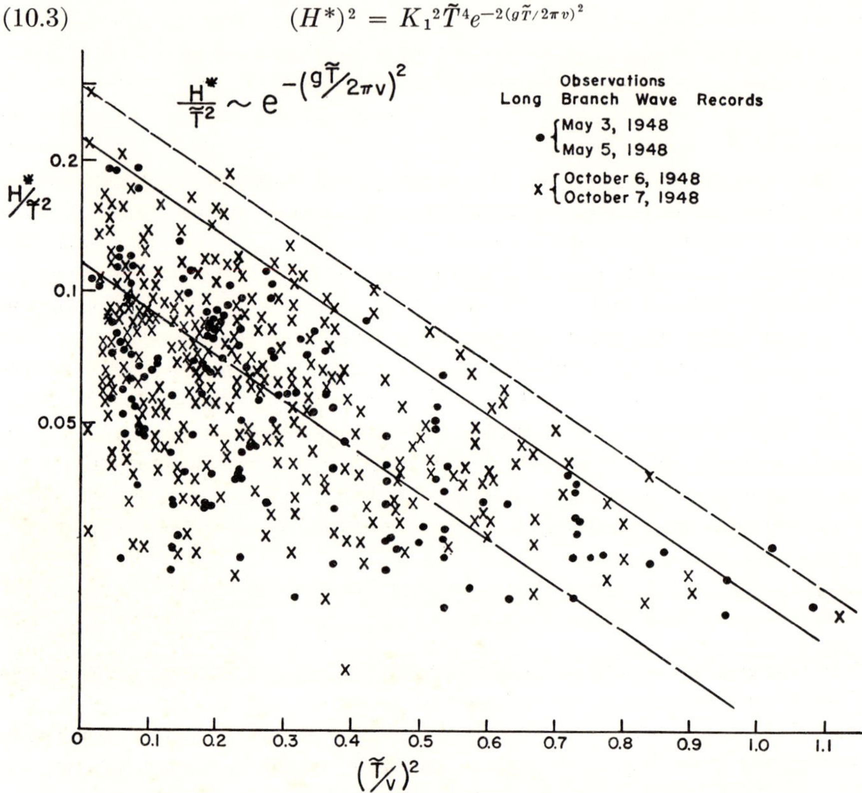

FIG. 4. Ratio of wave heights to the square of the apparent wave periods plotted against the square of the ratio of the apparent wave period to the wind velocity (after Neumann).

The above result applies to a small but finite band of the *periods*. If the band could be made infinitesimally wide, $K_1{}^2$ would become infinitesimal.

The area under the energy spectrum, $[A(\mu)]^2$, over an interval $d\mu$, is directly proportional to the square of this characteristic height, and, on factoring $K_1{}^2$ into the product of an infinitesimal and a new constant, equation (10.3) can be written as

$$[A(\mu)]^2d\mu = C_1T^4e^{-2(gT/2\pi v)^2}dT \tag{10.4}$$

Since an infinitesimal band is considered in (10.3), the symbol T is used instead of the symbol $\tilde{T}$ because true spectral periods are now involved. Since $\mu = 2\pi/T$ and since $dT = -2\pi\mu^{-2}d\mu$, equation (10.4) can finally be written as

$$[A(\mu)]^2 d\mu = \frac{\pi}{2} \frac{C}{\mu^6} e^{-2g^2/\mu^2 v^2} d\mu \tag{10.5}$$

Note that the minus sign becomes a plus sign because of the fact that a positive change in μ requires a negative change in T.

10.3. The Wave Heights

For a fully developed sea at constant v, all spectral frequencies from zero to infinity should be present. The total value of E is thus

$$E = \int_0^\infty \frac{\pi}{2} \frac{C}{\mu^6} e^{-2g^2/\mu^2 v^2} d\mu = \frac{C\pi \sqrt{\pi}\, 3v^5}{\sqrt{2}\, 2^6 g^5} \tag{10.6}$$

From equation (9.1) and from the results of Longuet-Higgins [49], Barber [48], and Watters [51], the average height of the $\frac{1}{10}$ highest waves is given by

$$\bar{H}_{1/10} = 3.60\sqrt{E} = 3.60 \left[\frac{\pi\sqrt{\pi}\, 3C}{\sqrt{2}\, 2^6 g^5} \right]^{1/2} v^{2.5} \tag{10.7}$$

If the spectrum derived by Neumann is correct, the significant height of the waves in a fully developed sea should be proportional to the wind velocity to the 2.5 power since the significant height equals 0.786 times the average height of the $\frac{1}{10}$ highest waves. This is in conflict with earlier results of other authors who reported that the significant height was proportional to the wind velocity to the second power or even to the 1.5 power.

Figure 5 shows reasons for believing that the exponent of v should be 2.5. The slope of the line given by equation (10.8) is $2\frac{1}{2}$ to 1 in Fig. 5. In (10.8), $H_{1/10}$ is in centimeters and v is in centimeters per second

$$\bar{H}_{1/10} = 0.9 \times 10^{-5} v^{2.5} \tag{10.8}$$

Neumann states that this result needs further substantiation from information based on more data, but the fact that Fig. 5 gives results which support the theoretical spectrum given in equation (10.5) is most interesting.

This result is not as much in conflict with the results of Sverdrup and Munk [56] as it would first appear to be. The curves $\bar{H}_{1/3} = K_{SM}v^2$ and $\bar{H}_{1/3} = K_N v^{2.5}$ cross at a velocity of about 30 knots. Hence the significant heights of the waves in a fully developed sea for a wind less than about 30 knots is less in Neumann's theory than in the theory of Sverdrup and

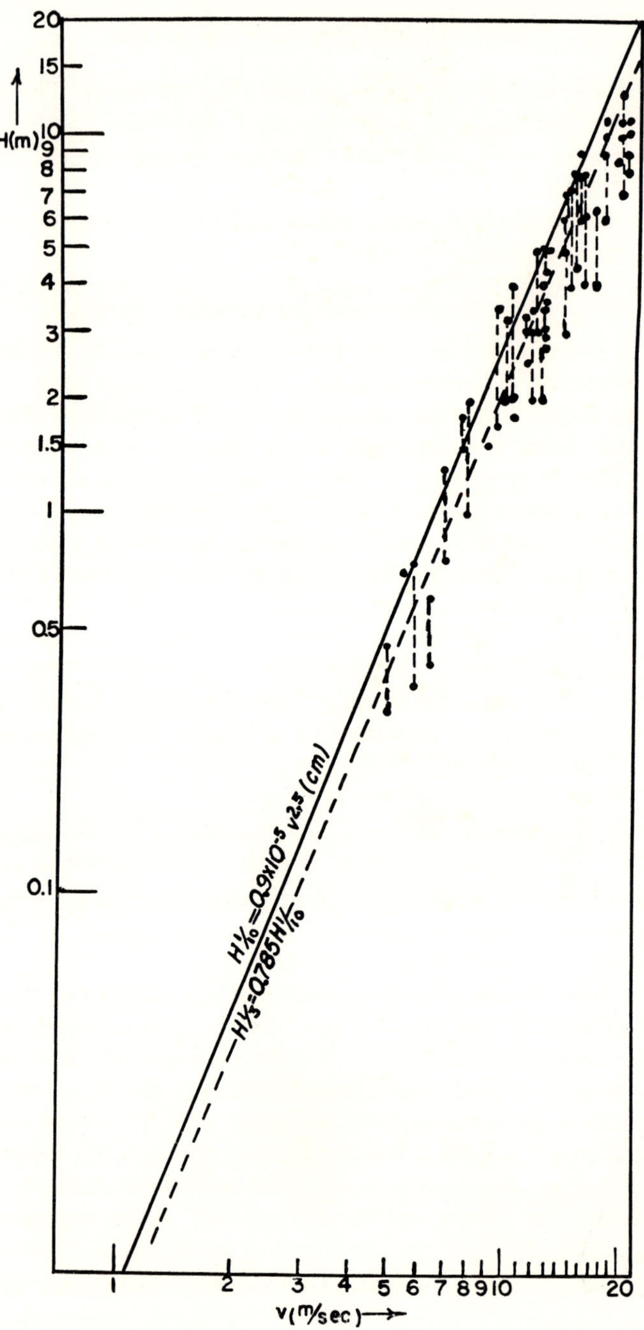

FIG. 5. Significant wave height and average of the one-tenth highest waves plotted gainst the wind velocity (after Neumann).

Munk, and it is higher for winds greater than 30 knots. The differences are small over the range of most reported significant heights, and final verification must await better data on wave heights for the higher wind speeds.

From (10.7) and (10.8), the constant, C, in (10.5) can be determined. It is equal to $3.05 \times 10^4 \text{ cm}^2 \text{ sec}^{-5}$.

10.4. The Shape of the Spectrum and the Various "Periods"

The spectrum for values of the frequency near zero is very low because of the effect of the exponential term. The requirement that $\mu_i = g/2v$ is another way to state that the group velocity of waves with the frequency μ_i equals the wind velocity, and when $[A(\mu)]^2$, as given by (10.5), is integrated from 0 to $g/2v$ it is easy to show that the energy present is only a minute fraction of the total energy. Thus the theoretical spectrum shows that no appreciable amount of energy associated with components with group velocities greater than or equal to the wind velocity is present in a fully developed sea.

The spectrum as a function of frequency begins to rise rapidly at a value of μ approximately equal to $g/1.6v$.

The maximum of the spectrum is found by differentiating $[A(\mu)]^2$ and solving for the frequency at which $[A(\mu)]^2$ is a maximum. The result is

$$\mu_{\text{max}} = \sqrt{2/3}\, g/v \tag{10.9}$$

This corresponds to a ratio of wave crest speed to wind velocity equal to $\sqrt{3/2}$ or 1.22. Hence the phase velocity of the spectral components with the greatest energy is slightly greater than the average wind velocity. This suggests that these components may be generated by the gusts in the turbulent wind field of the generating area.

On the open ocean, the only way to measure wave "periods" in the past has been by means of a stop watch. A stop watch observation is made by recording the time intervals between the successive crests as the waves pass a fixed point. Sometimes only the time intervals associated with the more dominant crests are recorded. However, as Neumann has shown, these "periods" are distributed over a wide range of values (from 5 to nearly 17 sec for a wind of 30 knots, for example) in a fully developed sea. The average of these "periods" is the average "period" as designated by $\bar{\bar{T}}$.

Equation (7.2) can be applied to equation (10.5), and the result is that $\bar{\bar{T}}$ is given by

$$\bar{\bar{T}} = \frac{\sqrt{3}\,\pi v}{g} \tag{10.10}$$

The frequency which corresponds to this "period" is given by

$$\bar{\bar{\mu}} = \frac{2}{\sqrt{3}}\frac{g}{v} = \frac{g}{0.866v} \tag{10.11}$$

These average "periods" for different wind velocities are quite close to the "periods" reported by some observers for fully developed seas. A particular average "period" for a given wind is actually much less than the corresponding period in the spectrum where the maximum energy occurs. Hence Cornish [57] reports that the average storm sea travels with a propagation velocity of about 80% of the wind velocity.

Cornish [57] made careful observations, but he did not know that he was observing a particular realization of a three dimensional stationary Gaussian process (and not a mildly disturbed simple harmonic progressive wave). He could also never have guessed that a formula developed in electronics in 1944 (Rice, equation (9.2)) would show that such stop watch observations were deceptive and misleading and that the maximum energy in the waves would be found at a completely different true spectral period.

In its original form, equation (9.2) as given by Rice gave the average number of zeros per second in a Gaussian noise. To obtain (9.2) this average number of zeros was cut in half, and the reciprocal was taken to express the results in terms of the average "period." Smaller oscillations which are superimposed on the waves, and which do not have sufficient range so that they pass through the mean elevation of the sea surface at the point of recording, are therefore not counted.

If these oscillations are counted, the result is essentially that the average number of maxima in the record is counted. From equation (9.3) it can be shown that the average time interval between the maxima is

$$\bar{\bar{T}}_{\text{maxima}} = \bar{\bar{T}}/\sqrt{3} \tag{10.12}$$

The occurrence of three different values for different kinds of "periods," equations (10.9), (10.10), and (10.12), explains the difficulty in interpreting past visual observations where the "period" assigned to the fully developed seas was associated with a ratio of phase speed to wind velocity which varied all the way from 1.4 to 0.8. Neumann [39] has shown that equation (10.10) agrees remarkably well with actually observed average "periods."

10.5. The Growth of the Waves

As a wind of constant mean speed and direction begins to blow over a stretch of water from an initial state of calm, the waves begin to grow in height, and the "period" of the waves increases gradually with time.

A steady state is attained, as a function of distance from the rear edge of the area over which the wind blows, at duration times which increase as the distance given above increases. At a certain distance from the rear of a fetch and after sufficient lapse of time, the fully developed state of the sea as defined by the complete spectrum given in (10.5) is attained.

For shorter durations or fetches, the complete spectrum is not present. When the complete spectrum is not present, Neumann [36a, b, 37] has shown that, for a given wind and a given duration or fetch, the energy in the spectrum given by equation (10.5) is present *above* a certain frequency, which is a function of either the duration or fetch, and that little or no energy is present below this frequency.

The two most important cases have been considered. The first case considers a constant wind with mean velocity v which blows over an unlimited stretch of water (the fetch, F). In this case, the energy added to the complex sea is the same everywhere, so that the waves grow at all localities at the same rate with time. In this case, then, the stage of development of the spectrum depends only on the duration of the wind.

The second case considers a wind which has blown for an infinitely long time but such that the stretch of water over which the wind blows is limited at the upwind end, for example, by the presence of land. Under these conditions the state of development of the spectrum depends only on the distance from the point at which the wind first affects the sea surface.

For the combined case in which the winds starts to blow at time zero over a limited fetch, the spectrum first grows according to the duration of the wind at a fixed point until the length of the fetch for that point limits further growth. Then at that time a steady state determined by the length of the fetch is obtained.

If the duration exceeds a certain length of time and the fetch exceeds a certain distance, the fully developed spectrum is attained. For example, all points downwind from a point 10 nautical miles offshore from some land would have the fully developed spectrum after 2.4 hours, if the wind had a velocity of 10 knots. For a 20-knot wind, the point would have to be 75 nautical miles offshore and the wind would have had to blow for 10 hours. For a 30-knot wind, the corresponding values are 280 nautical miles and 23 hours; for 40 knots, they are 710 nautical miles and 42 hours; and for 50 knots, they are 1420 nautical miles and 69 hours. Thus for winds of 30 knots or so the fully developed state is often obtained over large areas. For winds of 50 knots, areas of full development would be rare.

A frequency, μ_i, called the frequency of intersection, can be defined which is a function of the wind speed and either the duration or the fetch such that equation (10.13) holds

$$\mu_i = \begin{cases} \mu_i(v,F) & \text{if} \quad E_F < E_t \\ \mu_i(v,t) & \text{if} \quad E_t < E_F \end{cases} \tag{10.13}$$

The spectrum of the waves is then given by

$$[A(\mu)]^2 = \begin{cases} \frac{\pi}{2}\frac{C}{\mu^6} e^{-2g^2/\mu^2v^2} & \text{if} \quad \mu_i < \mu < \infty \\ 0 & \text{otherwise} \end{cases} \tag{10.14}$$

Hence the value of E at a given point depends on either the duration or the length of the fetch, and it is given by

$$E = \int_{\mu_i}^{\infty} [A(\mu)]^2 d\mu \tag{10.15}$$

where $E = E_t$ if the spectrum is controlled by duration and $E = E_F$ when and if the spectrum becomes controlled by the length of the fetch.

To obtain the theory of the generation of the waves, Neumann divided the spectrum into three broad spectral bands and approximated the energy in each band by a "characteristic wave." It was assumed that the spectrum grew from high frequency to low frequency by passing through these three spectral bands to the point where the sea was fully generated.

At first, the short-period wave components develop which remain in a quasi-steady state, always being regenerated after breaking by the energy supplied from the winds. After these waves have been fully generated, higher-period wave trains are generated which cover a spectral band around a "characteristic wave" with a spectral frequency such that the phase velocity at that frequency equals the wind speed. After the energy in this band of frequencies has been supplied to the developing sea, the wave components at the low-frequency end of the spectrum can build up if the duration and the fetch are long enough. These waves have phase velocities greater than the wind velocity and, in the earlier work by Neumann, it was found that the "characteristic wave" for the long-period waves had a ratio of phase velocity to wind speed of 1.19. This corresponds remarkably well to the value 1.22 obtained from equation (10.9).

The actual equations for the generation of the waves are too extensive to reproduce in this paper, but they depend on the ratio of the phase speed to the wind velocity and on the difference between the energy supplied by the wind to the waves and the energy dissipated in the complex wave motion. When the energy supplied equals the energy dissipated, the steady state is obtained.

Instead of being exactly vertical at the frequency of intersection μ_i (equation (10.13)) the spectrum of a partially developed sea may actually begin to rise steeply, but not vertically, at a value of slightly less than μ_i, and it may not attain full development until μ is a little larger than μ_i. Thus "periods" observed by stop watch associated with frequencies equal to $0.85\mu_i$ are observed occasionally.

Neumann [39] has plotted many histograms of the distribution of the "periods" obtained by stop watch when observing the waves at a fixed point. For cases of limited fetch and duration, it was found that there were "periods" present up to a certain upper bound and that above this value none were present. From a study of such histograms, Neumann was able to extend the original concept of three "characteristic waves" to the concept of a continuous spectrum and to define the frequency of intersection. From this information the rise at μ_i appears to be quite steep. Also, early spectra analyzed by Klebba [4], and unpublished analyses of some spectra analyzed by the Tukey method, confirm this part of Neumann's theory.

If the family of curves given by

$$E(\mu,v) = \int_{\mu}^{\infty} \frac{\pi}{2}\frac{C}{\omega^6} e^{-2g^2/\omega^2v^2} d\omega \tag{10.16}$$

is plotted against the parameter v, the result is a set of co-cumulative spectra. When $\mu = 0$ the value given by (10.6) is attained. The curve is most steep at a frequency given by (10.9). It then gradually becomes asymptotic to zero as μ approaches infinity. The difference between two E values at two frequencies on a curve for constant v is a measure of the potential energy (apart from a constant) in the wave motion due to these frequencies.

The lines given by (10.13) can then be plotted on the co-cumulative spectra. The intersection of a fetch line with a co-cumulative spectrum for a particular v then determines μ_i, E_F, and the frequencies present in that particular partially developed sea, since only those frequencies to the right of μ_i are present. The energy present in any particular band can also be determined.

Moreover, as the wind continues to blow over the fetch, the significant wave height ($2.83\sqrt{E}$) will continue to grow until either the length of the fetch limits further growth or until the fully developed state is attained after a certain duration of the wind. Also, by virtue of equation (9.2), the average "period" of the waves will continuously increase until the average "period" of the fully developed sea is attained, if the fetch is long enough and the wind blows long enough.

The true spectral periods are either present or they are not present, but the average "period" changes with time. Thus the paradoxical expression $(1/T)(dT/dt)$, which makes no sense when applied to the *period* of a simple harmonic progressive wave, has a meaning when it is written as $(1/\bar{\bar{T}})(d\bar{\bar{T}}/dt)$. However, the expressions which involved notations like $(1/T)(dT/dt)$ were derived under the assumption that the wave crests were conservative and that T was a slowly varying function of time and space. Since the crests of a short crested Gaussian sea surface are not conservative, and since the individual "periods" vary widely at a given point, the possible meaning of such equations must be re-evaluated.

11. The Angular Variation of the Spectrum

The spectrum of the sea surface is really a function of two variables. The wind generated sea is visibly short-crested. Areas of the sea surface can be found on which trains of high and well-defined waves exist, but these waves can never be followed very far along the crests before they become low and merge into an area where the crests are not well defined. Short-crestedness is an essential property of all waves. Ways must be developed to measure this short-crestedness accurately. The statistical methods given above and the theoretical spectrum of Neumann apply only to waves observed as a function of time at a fixed point. The statistical properties of the short crested sea surface, such as the length of crests along the crests, and the probability distribution function of the highest elevation of each short crested wave, are not understood.

Let positive x be the direction toward which the wind is blowing in the representation of the short crested Gaussian sea surface, and assume that no swell is present. Then the spectrum of the waves should probably be confined to an angular variation from $-\pi/2$ radians to $+\pi/2$ radians since it is difficult to see how a wind can generate a disturbance which travels in a direction opposite to the wind.

From equations (6.20) and (10.5), one can obtain

$$\int_{-\pi/2}^{\pi/2} [A(\mu,\theta)]^2 d\theta = \frac{\pi}{2} \frac{C}{\mu^6} e^{-2g^2/\mu^2 v^2} \tag{11.1}$$

There are an infinite number of functions which can satisfy this relationship. The two-dimensional spectrum could be concentrated over a narrow angular band about plus and minus θ, and it would then be quite high; or it could be spread out quite low. The integral over θ would still give the same result. Measurements have not been made which would give information on this angular variation.

In wave forecasting methods used during the past five or ten years, the rule for forecasting swell arriving from a distant storm was that swell should be forecasted if the point of the forecast was at an angle of

± 30 degrees to the direction of the wind over the storm. Thus if a generating area were 400 nautical miles wide and if a forecast point was 800 nautical miles away at a point at an angle of 30 degrees to the center of the forward edge of the generating area, the swell would be forecasted to be equal in height to the swell 800 nautical miles directly in front of the storm. Yet the point at an angle would be 200 miles in the shadow of the envelope described by equation (4.19), if all spectral components were traveling in the direction of the wind.

The only conclusion is that the spectral components in the sea have appreciable energy at 30 degrees to the wind. Hence $[A(\mu,\theta)]^2$ must be quite high over this range of θ. More careful analyses of many such wave situations show that waves even arrive at angles of plus and minus 45 degrees to the wind in the storm. There are other cases in which waves arrive at forecast points at even greater angles to the winds.

Arthur [58] has studied this problem, and he found that the wave height varied like the cosine of the angle of the forecast point to the wind. Hence, if the wave height equals 1 directly in front of the generation area, it should be 0.866 at 30 degrees and 0.707 at 45 degrees.

An estimate of the functional form of $[A(\mu,\theta)]^2$ for any sea can be made on the basis of these hints as to its form. In cgs units, this estimate is

$$[A(\mu,\theta)]^2 = \begin{cases} \dfrac{Ce^{-2g^2/\mu^2v^2}}{\mu^6}(\cos\theta)^2 & \text{for } \mu_i < \mu < \infty \\ & \text{and for } -\dfrac{\pi}{2} < \theta < \dfrac{\pi}{2} \\ 0 & \text{otherwise} \end{cases} \tag{11.2}$$

This spectrum produces forecasts of swell which agree with current observational knowledge as will be shown later. The wave forecasting methods given by Pierson, Neumann, and James [50] use this spectrum in the theory of swell forecasting. It appears to work fairly well.

The spectrum given in equation (11.2) still need not be correct. One way to really find out is to measure the short crested Gaussian sea surface from stereo-aerial photographs and compute the function $[A(\mu,\theta)]^2$. It is possible that the range of θ will be narrower for low values of μ and wider for high values of μ than equation (11.2) suggests.

Such stereo-aerial photographs were recently obtained through the joint efforts of the Office of Naval Research, the United States Navy, The U. S. Hydrographic Office, David Taylor Model Basin, and Woods Hole Oceanographic Institution.

Over 100 stereo pairs each covering an area of the sea surface approximately 2000 ft by 1000 ft were obtained. Two aircraft approximately 2000 feet apart were used, and cameras pointed straight down in each aircraft were simultaneously triggered by means of an FM radio link. These data are to be analyzed by the Hydrographic Office, and the results will help to solve many of the problems discussed in this text. Papers on the results of the analysis should be available in about a year.

The work of coordinating and planning the project was carried out by R. C. Vetter of the Office of Naval Research and Wilbur Marks of the Woods Hole Oceanographic Institution.

However, one interesting result can be obtained which suggests that equation (11.2) may not be too far off. From the results of section 6.4, the spectrum $[A(\nu_0;0)]^2$ for a fully developed sea can be evaluated. When this spectrum is substituted into equation (9.4), Pierson [46] has shown that, with the aid of equation (10.10), one can obtain

$$\bar{\bar{L}} = \tfrac{2}{3} g \bar{\bar{T}}^2 / 2\pi \tag{11.3}$$

Hence the average "wavelength" in a fully developed sea is *not* obtained by the use of the average "period" in the classical formula. Observations described by Deardorf [59] and discussions with operational wave forecasters confirm that the above formula is more nearly correct in a sea. In a swell, the classical formula is nearly correct, but it must always be remembered that ocean waves are not simple harmonic progressive waves.

Short crested waves are difficult to generate in wave tanks, but long crested Gaussian seas have been generated by Lewis.* Such waves could be represented as a function of x and t by setting $\cos\theta$ equal to 1 and $\sin\theta$ equal to 0 in equation (6.1). Then the two variable spectrum could be immediately integrated over θ to obtain the Neumann spectrum as a function of μ alone. The waves generated by Lewis were Gaussian with a spectrum very much like one which would result from a scaled down Neumann spectrum. Equations similar to equations (9.2) and (9.4) were then applied to the spectrum (as determined from the equations given in subsection 13.3). The observed average "period" was 1.1 sec and the computed value was 1.07 sec. The observed average "wavelength" was 5.2 ft and the computed value was 5.4 ft. The classical formula applied to the average "period" yields a value of 6.19 ft.

If an attempt is made to compute the value of the average distance between successive maxima in a fully developed sea from equation (9.5), the attempt gives the result that $\bar{\bar{L}}_{\max} = 0$ since the denominator becomes infinite. The high frequencies cause the difficulty, and hence the spectrum derived by Neumann predicts that a fully developed sea is completely covered by ripples of very short wavelength. The formula for the spectrum

* Lewis, E. V. (1954). Ship model tests to determine bending moments in waves. Transactions of the Society of Naval Architects and Marine Engineers Vol. 62.

should be expected to fail at the frequencies of capillary waves anyway, and hence the above result is quite realistic.

Actual "wavelengths" are rarely measured on the actual sea surface. The "wavelengths" are usually computed from the "periods" with the aid of the classical formula. Hence they are usually wrong, since well defined swell is rare.

When the fact that partially developed seas, dead seas, and swells are not distinguished in current observation methods is considered, it is not at all surprising that reported average "wavelengths" range from the theoretical value computed from the classical formula to one-half of the theoretical value, and even to a tenth or less of the theoretical value, depending on the amount of ripple included. The sea surface is far more irregular as a function of x along a line at an instant of time then it is a function of time at a fixed point in space.

The spectra described above are slowly varying functions of space and time. The process which they represent is not, therefore, a stationary Gaussian process. However, depending on the meteorological conditions, areas of many hundreds of square miles for many hours, and wave records as a function of time lasting from twenty minutes to many hours, are so closely represented by treating them as samples from a particular realization of a stationary Gaussian process that the difference is unimportant. The term quasi-stationary Gaussian process describes such a system since the results of the theorem proved in section 6 are not valid if the translations x^*, y^*, and t^*, are too big.

12. Sea Surface Glitter and Wave Slopes

12.1. Reflection of the Sun by the Sea Surface

On a clear day, if the surface of the sea were absolutely mirror-like, an observer looking down would see the image of the sun reflected perfectly in the water. As soon as ripples and small waves develop, this image is broken up into many bright pin-points of light, and a dancing fluctuating glitter pattern is formed.

An example of such a glitter pattern is shown in Fig. 6 which is taken from Plate 8 of Part I of the paper by Cox and Munk [60].* This figure also shows an excellent example of the appearance of the short crested Gaussian sea surface. The wind was 27.2 knots at a height of 41 feet above the water. Notice the white caps.

If the sea surface is horizontal at a point where the sun's image would have appeared, a glitter would be returned to the observer from that

* See also Cox, C. and Munk, W. H., Statistics of the Sea Surface Derived from Sun Glitter. *J. Marine Research* (in press).

point. At points more distant from the position where the sun's image should be, the slope of the sea surface must be more and more steep in order to return a glitter to the observer. Thus, in Fig. 6, the glitters

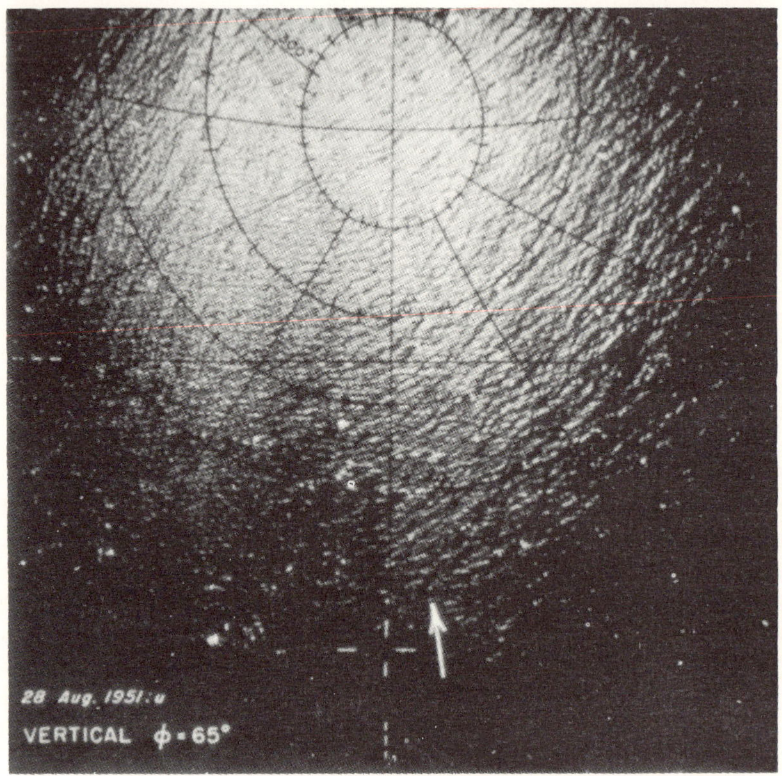

FIG. 6. Glitter pattern (after Cox and Munk).

become fewer and fewer the further away they are from the center of maximum concentration.

12.2. Distribution of Slopes as Found by Cox and Munk

Cox and Munk [60] analyzed the glitter patterns of 29 photographs like Fig. 6. They found that the slopes of the sea surface could be represented by a Gram-Charlier series. This series generalizes the Gaussian distribution by multiplying the bivariate Gaussian distribution by a series of Hermite polynomials in order to correct for the effects of skewness and kurtosis.

The distribution of crosswind slopes was found to be somewhat more peaked than the Gaussian distribution, and the distribution of the upwind-downwind slopes was found to be skewed so that the mode was not at zero slope. The nearness to the Gaussian distribution is quite evident, and the correction terms are quite possibly due to the superposition of nonlinear effects.

The variances of the cross wind (σ_c^2) and upwind (σ_u^2) slope components were found to be given by equations (12.1) and (12.2), where the wind is in meters per second as measured at 41 feet above sea level and where the data apply for clear sea surfaces with no slicks present. Equation (12.3) gives the sum of the cross wind and up wind variances. It is not equal to the sum of (12.1) and (12.2) because a least-squares fit was made to the individual observed values.

$$\sigma_c^2 = 0.003 + 1.92 \times 10^{-3}v \pm 0.002 \tag{12.1}$$
$$\sigma_u^2 = 0.000 + 3.16 \times 10^{-3}v \pm 0.004 \tag{12.2}$$
$$\sigma_c^2 + \sigma_u^2 = 0.003 + 5.12 \times 10^{-3}v \pm 0.004 \tag{12.3}$$

12.3. Effect of Slicks

If oil covers the sea surface, its effect is to damp out the high-frequency capillary waves much more than the gravity wave components at frequencies less than 4π. Both natural and artificial slicks were studied by Cox and Munk. The artificial slicks were made by pumping a mixture of 40% crank case oil, 40% diesel oil, and 20% fish oil onto the water. When slicks were present, the variances, as defined above, were given by equations (12.4) and (12.5).

$$\sigma_c^2 = 0.003 + 0.84 \times 10^{-3}v \pm 0.002 \quad \text{(slick)} \tag{12.4}$$
$$\sigma_u^2 = 0.005 + 0.78 \times 10^{-3}v \pm 0.002 \quad \text{(slick)} \tag{12.5}$$
$$\sigma_c^2 + \sigma_u^2 = 0.008 + 1.56 \times 10^{-3}v \pm 0.004 \quad \text{(slick)} \tag{12.6}$$

12.4. Comparison with Neumann's Spectrum

The free surface, $\eta(x,y,t)$, as given by equation (6.1) (or in a partial sum by equation (6.3)), can be differentiated with respect to x so that $\eta_x(x,y,t)$ is known. This new function will be a three-dimensional stationary Gaussian process. If equation (11.2) is the sea-surface spectrum (for $\mu_i = 0$), the spectrum of this process, which is a function which describes the slope of the sea surface in the x-direction as a function of space and time, is given by

$$[S_x(\mu,\theta)]^2 = \begin{cases} \dfrac{C(\cos\theta)^4}{\mu^2 g^2} e^{-2g^2/\mu^2 v^2} & \text{for} \quad -\dfrac{\pi}{2} < \theta < \dfrac{\pi}{2} \\ 0 \quad \text{otherwise} \end{cases} \tag{12.7}$$

The spectrum of the function which describes the slope in the y-direction as a function of space and time is given by

$$(12.8)\quad [S_y(\mu,\theta)]^2 = \begin{cases} \dfrac{C(\cos\theta\sin\theta)^2}{\mu^2 g^2} e^{-2g^2/\mu^2 v^2} & \text{for} \quad -\dfrac{\pi}{2} < \theta < \dfrac{\pi}{2} \\ 0 \quad \text{otherwise} & \end{cases}$$

These slopes are normally distributed in the sense of the ergodic theorem since a sample of many readings will be Gaussian. The variance of the slope in the x-direction is given by equation (12.9), and in the y-direction it is given by equation (12.10), where v is in meters per second, but where the wind is measured at anemometer level

$$(12.9)\qquad \sigma_x^2 = \int_0^\infty \int_{-\pi/2}^{\pi/2} \frac{[S_x(\mu,\theta)]^2}{2}\, d\theta d\mu = 1.19 \times 10^{-3} v$$

$$(12.10)\qquad \sigma_y^2 = \int_0^\infty \int_{-\pi/2}^{\pi/2} \frac{[S_y(\mu,\theta)]^2}{2}\, d\theta d\mu = 0.40 \times 10^{-3} v$$

$$(12.11)\qquad \sigma_x^2 + \sigma_y^2 = 1.59 \times 10^{-3} v$$

The maximum in the slope spectra given by (12.7) and (12.8) as a function of frequency is

$$(12.12)\qquad \mu_{s,\max} = \sqrt{2}\, g/v$$

At a wind of 10 meters/sec, this corresponds to a period near 4 sec, and at 5 meters/sec, to a period near 2 sec. The area under the spectrum at a wind of 10 meters/sec for components with periods less than ½ sec is about 6% of the total area. The Neumann spectrum and the integral representation given in section 6 are not adequate at high frequencies corresponding to capillary waves.

12.5. Discussion of Results

If the major effect of the presence of a slick is to remove from the spectrum those components corresponding to capillary waves, and if Neumann's spectrum represents the gravity wave contribution to the sea surface slopes, then equations (12.6) and (12.11) can be compared judiciously. The remarkable agreement is probably in part fortuitous, since the exact form of the spectrum at high frequencies has not as yet been actually measured.

The various constants and parameters which determined the Neumann spectrum were measured from the properties of the dominant waves in the sea. The winds were measured at heights from 20 to 30 feet, and the spectrum is most nearly applicable if the wind is measured within this height range and if the sea-surface temperature is warmer than the air temperature. The winds used in the equations obtained by

Cox and Munk were measured at 41 feet, and the sea temperature was slightly cooler than the air temperature.

If the constants in these two equations are varied by ± 10 to 20%, the range in the values would then probably include these factors, and the agreement would then still be good.

The value of σ_x^2 is three times larger than the value of σ_y^2, whereas the values of σ_c^2 and σ_u^2 are practically equal. The term $\cos^2 \theta$ in (11.2) accounts for the value of the theoretical ratio. The observations show that the θ variability at high frequencies is even greater than would be indicated by section 11.

However, since the high-frequency part of the spectrum is emphasized in the slope pattern, this result does not seem to be an adequate reason for discarding equation (11.2) in a working model for practical wave forecasting. Further study should eventually answer this question.

In contrast with these results, Barber* in some recent unpublished work has constructed a device which measures the variation in θ of $[A(\mu, \theta)]^2$ for a fixed value of μ. The method depends upon the analysis of the correlation function instead of the analysis of the phase differences for the different directions, and it gives considerable information on the basis of data from an ingeniously oriented array of wave recorders.

Some preliminary measurements were made on wind waves in Waitemata Harbour, Aukland, N.Z., for frequencies corresponding to waves with a period of 1.85 sec. The results gave a curve quite similar in shape to $(\cos \theta)^2$ except that it was somewhat narrower. All of the energy appeared to be confined to an angular range of ± 60 degrees.

It should be pointed out, as Cox and Munk have done, that many other forms of theoretical energy spectra could be derived which would yield this linear variation of σ_x^2 and σ_y^2 with the wind velocity. Thus, final proof of the validity of the Neumann spectrum must await the actual determination of some spectra.

The theoretical derivation of Neumann and the excellent and carefully detailed study of Cox and Munk were carried out independently and nearly simultaneously. Both results were presented during the sessions of the American Geophysical Union in the Spring of 1953. The facts that the spectrum of the waves predicts a variance of the slopes which is proportional to the wind velocity, that this linear variation is observed, and that a derivative of the representation given by (6.1) still is approximately Gaussian (derivatives of such functions behave more erratically than the functions themselves) all suggest that the theory of wind generated gravity waves is now on a firm theoretical foundation.

* Barber, N. F. (1954). The direction of travel of sea waves. Dept. of Scientific and Industrial Research, Geophysics Division, Wellington. (Unpublished manuscript.)

13. Wave Record Analysis

13.1. Comparison of Numerical and Electronic Methods

There are two ways to analyze a wave record in order to find its spectrum. One way is numerical. The other is to use an electronic analyzer. Both methods depend on the same basic statistical concepts.

One numerical way would be to make a Fourier series analysis of a finite section of a particular realization. The Fourier coefficients, A_n and B_n, are shown by Rice [33] to be normally distributed with variances related very nearly to the true energy in the band which extends from $2\pi(n - \frac{1}{2})/\tau$ to $2\pi(n + \frac{1}{2})/\tau$ over the true *but unknown* continuous spectrum. The sum of the squares of the Fourier coefficients

$$C_n^2 = A_n^2 + B_n^2$$

is distributed according to $e^{-x/\Delta E}$ for positive x where ΔE is *nearly* equal to the true energy in the band.

13.2. Electronic Analysis

The Admiralty Research Laboratory analyzer, described by Barber and Ursell [60] and Deacon [3], found the values of C_n as described above.* The spectrum was therefore quite erratic. The analyzer described by Klebba [4] probably smoothed over one or two values of C_n and failed to average over a complete cycle of the record. Hence it did not give discrete spikes in an analysis. Also, since it looked at different parts of the record during each analysis, it gave different spectra for repeated analyses of the same record.

The energy spectrum of a wave record can best be recovered in its approximate shape by averaging electronically over a band of values of C_n^2 with a filter with an appropriate shape. Such filters have been designed by Chang,† and an analyzer with a nearly perfect design filter has been constructed by him. Confidence limits can then be assigned to the smooth spectral curve on the basis of a Chi-square distribution with the number of degrees of freedom determined by the band width and shape of the filter. For a study which shows practical ways to use the analyzer in problems connected with beach erosion, see the Technical Memorandum cited below.‡

* Personal communication from Robin A. Wooding.

† S. S. L. Chang (1954). "On the filter problem of the power spectrum analyzer," *Proc. I.R.E.* **43,** No. 8.

‡ Pierson, W. J., Jr. (1954). An electronic wave spectrum analyzer and its use in engineering problems. Technical Memorandum No. 56, Beach Erosion Board, Washington, D. C.

13.3. The Tukey Method

Methods devised by Tukey [34] eliminate the difficulty which would occur if an attempt to make an actual numerical analysis by Fourier series methods were made. By forming lagged products of a series of discrete values read off the record, the covariance function is approximated. The Fourier series cosine analysis of this function then gives estimates of the energy in bands of the spectrum which include many values of C_n^2. A smoothing function then gives the best estimate of that part of the total variance of the record (U_h) which is associated with frequencies between $2\pi(h - \frac{1}{2})/2\Delta tm$ and $2\pi(h + \frac{1}{2})2\Delta tm$ where Δt is the spacing of the points in the original sample and m is the number of lagged products computed.

The formulas due to Tukey [34] are given for ready reference in equations (13.1)–(13.4) where the values $\eta(t_1)$, $\eta(t_2)$, $\eta(t_3)$, . . . , $\eta(t_N)$ are spaced Δt units apart

$$Q_p = \frac{2}{N - p} \sum_{k=1}^{N-p} \eta(t_k)\eta(t_{k+p}) \qquad p = 0, 1, \cdots, m \tag{13.1}$$

$$L_h = \frac{1}{m}\left[Q_0 + 2 \sum_{p=1}^{m-1} Q_p \cos \frac{\pi ph}{m} + Q_m \cos \pi h\right] \qquad h = 0, 1, \cdots, m \tag{13.2}$$

$$U_h = 0.23L_{h-1} + 0.54L_h + 0.23L_{h+1} \tag{13.3}$$

In (13.3), $L_{-1} = L_{+1}$, and $L_{m+1} = L_{m-1}$.

The numbers U_h are distributed according to a Chi-square distribution with f degrees of freedom such that the expected value of U_h is the true energy in the band.

$$f = \frac{N - m/4}{m/2} \tag{13.4}$$

Consequently, any desired confidence limits, usually 90%, can be assigned to each computed value. For f large, which means a large N, a small m, relatively, and lots of work, the confidence limits bound the values of U_h closely and the shape of the spectrum can be seen. For a fixed N, high resolution, large m, requires a sacrifice of precision; and high precision, large f, requires a sacrifice of resolution.

The value of Δt must be judiciously chosen. If the period $2\Delta t$ is not small enough so that energy at periods less than $2\Delta t$ is present, energy is falsely reported, or aliased, into lower frequencies, effectively, by folding the true spectrum back on itself at the frequency $2\pi/2\Delta t$ and adding these contributions to the lower frequencies. If Δt is too small, zero values are reported at the high-frequency end, and needless work is done. The rule is to sample at twice the frequency of the lowest period present.

13.4. Pressure Wave Records

For pressure wave records, the value of Δt is determined from the depth in which the pressure recorder is located. Twice that depth in feet is the wavelength in feet of the deep-water wave whose height would hardly be recorded by the recorder. One-half of the square root of the number which results from dividing this wavelength by five is then equal to the value of Δt. Thus Δt should be 1.5 sec if the depth is 22.5 feet. For further information on the application of the results of Tukey to the analysis of wave records, see Pierson and Marks [44].

Most of the information lost by the filtering effect of depth can be regained from the spectrum of the pressure record by multiplying it by

$$\left[\cosh\left(\frac{\mu^2 h}{g}\, I\left(\frac{\mu^2 h}{g}\right)\right)\right]^2$$

Only the energy for periods less than $2\Delta t$ is then lost.

14. Ship Motions

A ship moving with a constant heading and a fixed speed through the short crested Gaussian sea surface described in section 6 responds with rolling, pitching, and heaving motions which themselves are particular realizations of quasi-stationary Guassian processes as a function of time if the amplitudes of the motions are not too great. The spectrum of the ship response can be obtained from the spectrum of the waves and the properties of the ship. From the motion spectrum, the statistical properties of the motion can be predicted since the results of Rice [33], Wooding, and Longuet-Higgins [49] are equally applicable to ship motion records.

The theory of the motions of a ship in confused seas is given by St. Denis and Pierson [8]. The response amplitude operators of the ship are derived theoretically, the product of this operator and the spectrum of the waves, $[A(\mu,\theta)]^2$, is found, and the result is mapped by a transformation and an integration into a spectrum of one variable where this variable is the frequency of encounter of the ship with the waves.

Stoker and Peters [62] have just completed a rigorous derivation of the linear response of a ship to a simple harmonic progressive wave. The interaction of the ship and the water is treated, and hence the Froude-Krylov hypothesis is eliminated. Since the theory is linear it can also be applied to the motions produced by a short crested Gaussian sea surface as described in section 6.

15. Wave Refraction

15.1. Refraction of a Simple Harmonic Progressive Wave

When a simple harmonic progressive wave in deep water travels into an area where the depth is less than one-half of the deep-water wavelength, the waves are refracted. The works of Friedrichs [22], Peters [23], Eckart [24], and Arthur, Munk, and Isaacs [27] can be applied to determine the nature of the waves in the shallower water.

For application to irregular bottom contours, the procedures given by Arthur, Munk, and Isaacs are a practical method for determining the effect of refraction. After refraction the crests of a simple harmonic progressive wave are usually no longer straight and parallel. Hence the wave system in the refraction zone is no longer a stationary Gaussian process in three variables.

At a point of observation in the refraction zone, the period of the wave is the same, but the height is larger or smaller depending upon the effect of convergence or divergence of the wave rays and upon the group velocity appropriate for the depth of the water at that point.

Let $[K(\mu,\theta)]^2$ be a function which tells by how much the square of the amplitude of a simple harmonic progressive wave with a frequency μ, traveling in the direction θ in deep water, must be multiplied in order to obtain the square of the amplitude of the wave at a point of observation after refraction. Let $\theta_R = \Theta(\mu,\theta)$ be a function which tells the direction toward which the crest will be traveling at the point of observation after refraction if the wave had the direction θ and the frequency μ in deep water. Both of these functions can be estimated by numerical methods with the procedures given by Arthur, Munk, and Isaacs [27].

15.2. Refraction of Short Crested Gaussian Waves

The refraction of actual waves is studied by representing the waves in deep water by a partial sum such as equation (6.3) and operating on the partial sum by the functions described above. In the limit, the energy in the spectrum after refraction is found by multiplying $[A(\mu,\theta)]^2$, whether it be a fully arisen sea, a partially generated sea, or a swell spectrum (as defined in the next section), by $[K(\mu,\theta)]^2$. The new directions of the crests can be inserted into this product by inverting $\theta_R = \Theta(\mu,\theta)$ to find $\theta = \Theta^*(\mu,\theta_R)$, substituting the result into the above product, and multiplying by the Jacobian $\partial(\theta,\mu)/\partial\theta_R\partial\mu$. (Note that μ before refraction equals μ after refraction.) The spectrum after refraction is therefore given by

$$(15.1)\qquad [A_R(\mu,\theta_R)]^2 = [A(\mu,\Theta^*(\mu,\theta_R))]^2[K(\mu,\Theta^*(\mu,\theta_R))]^2\Gamma(\mu,\theta_R)$$

In equation (15.1), the function $\Gamma(\mu,\theta_R)$ comes from the Jacobian. It is given by

$$(15.2)\qquad \Gamma(\mu,\theta_R) = \partial\Theta^*(\mu,\theta_R)/\partial\theta_R$$

At the point of observation, the wave system can be approximated by

$$(15.3)\quad \eta(x_0,y_0,t) = \int_0^\infty \int_{-\pi}^{\pi} \cos\left[\frac{\mu^2}{g} I\left(\frac{\mu^2 h}{g}\right)[x_0 \cos\theta_R + y_0 \sin\theta_R] - \mu t + \epsilon(\mu,\theta_R)\right] \cdot \sqrt{[A_R(\mu,\theta_R)]^2 d\theta_R d\mu}$$

In equation (15.3), arbitrary translations in the x_0 and y_0 directions are not permitted because this representation in the x_0,y_0-plane is valid only in a limited region about the point of observation in the x_0,y_0-plane. This form, however, does suggest reasons why the crests of the waves are longer along the crests after refraction and why an echelon effect can often be observed.

As a function of time at a fixed point of observation, the process is still a stationary Gaussian process with a frequency spectrum $[A_R(\mu)]^2$ found by integrating out the angular variable θ_R. If a wave system is recorded by a pressure recorder (in terms of feet of static water pressure), the spectrum of the pressure oscillations is

$$(15.4)\qquad [A_{PR}(\mu)]^2 = [A_R(\mu)]^2 \Big/ \left[\cosh\left(\frac{\mu^2 h}{g} I\left(\frac{\mu^2 h}{g}\right)\right)\right]^2$$

Most wave records presently available are pressure records. Only a few spectra have been evaluated. Pressure records of refracted waves were what were studied by Rudnick [5], and nearly everyone else referred to here except Neumann, in studying wave properties. Spectra from such records must be corrected for the effect of depth and for the effect of refraction before they can be compared with Neumann's theoretical spectra and with the results given for deep-water waves. Also, the effects of bottom friction and percolation are not well enough understood to account for their effects quantitatively.

Refraction can modify the spectra markedly, as described by Pierson, Tuttel, and Woolley [63]. The results of the paper just mentioned were based on a guess as to the nature of the deep-water wave spectrum. Also, the applicability of equation (9.2) was not known at that time. Neumann's theoretical spectra had not yet been derived. If the study were repeated using a Neumann spectrum, the results would be even more striking. The height and the average "period" of the refracted waves would vary markedly at different points along the New Jersey coast.

16. The Propagation of Wind Generated Gravity Waves

16.1. General Remarks on Wave Propagation

Waves are not only generated by areas of high wind over the oceans, but they also travel for many thousands of miles in the form of swell. As soon as the forces of the wind are removed, the waves become free to travel according to the classical laws of gravity wave propagation. The only thing that needs to be done is to solve the initial value problem for the case where the waves in the generation area are described by a quasi-stationary Gaussian process.

16.2. An Initial Value Problem

Areas of high winds over the ocean frequently have the property that they last for several days and then die out in intensity. Suppose that such an area of high winds exists such that waves with a spectrum given by $[A(\mu),\theta)]^2$ over a rectangular area are produced and such that no waves are generated outside of this area. These requirements are practically impossible to fulfill exactly in nature, but every once in a while they are approximated by certain types of storms at sea.

In a rectangular coordinate system, at the time $t = 0$, define an area such that $-F < x < 0$ and $-W_s/2 < y < W_s/2$. Suppose that the waves which would be observed at time zero inside this area are those that would have existed if a stationary Gaussian process in three variables, as defined by equation (6.1), were present, and that outside of this area no disturbance at all would be observed. Let the spectrum $[A(\mu,\theta)]^2$ be some definite spectrum for either a fully developed or partially developed sea, as given by equation (11.2). The length, F, might be 500 nautical miles, and the width, W_s, might be 400 nautical miles for this area as it is shown in Fig. 7.

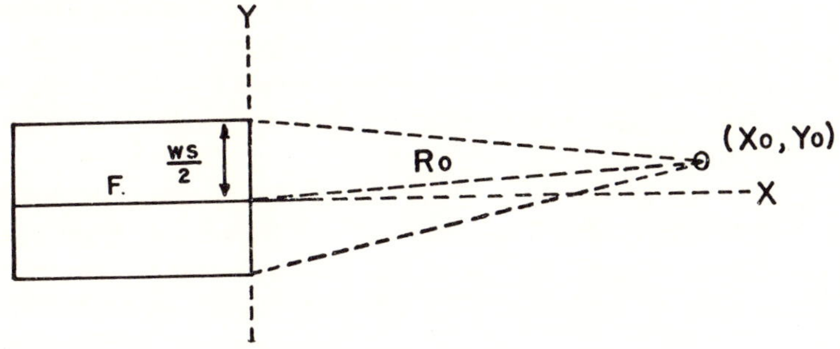

Fig. 7. The geometry of the initial value problem.

Assume also that no wind is blowing and that the effect of friction is negligible. Therefore, no more energy will be added to or removed from the system, and the spectrum will not change due to such an effect.

Near the center of this area, and at times after $t = 0$, translations of one or two hundred miles and of two or three hours would still yield the statistical properties discussed in previous sections. However, the system in some sense is traveling off into the positive half-plane.

Now put a raft at the point $x = x_0$, $y = y_0$, where x_0 is positive and y_0 is not too large compared to $W_s/2$. What will happen to the raft?

The partial sum given by equation (6.3) can be used to represent a particular realization of the sea surface given by a member of the ensemble described by equation (6.1). An arbitrarily small, but still finite, net could be used over the partial sum. A particular term in the sum of many terms can be singled out. The initial value problem of this particular term is, at $t = 0$, given by

$$(16.1) \quad \eta(x,y) = \begin{cases} A_{mq} \cos\left(\frac{\mu^2{}_{2m+1}}{g}(x \cos \theta_{2q+1} + y \sin \theta_{2q+1}) + \epsilon^*\right) \\ \quad \text{for} \quad -F < x < 0 \\ \quad \text{and} \quad -\frac{W_s}{2} < y < \frac{W_s}{2} \\ 0 \quad \text{otherwise} \end{cases}$$

where ϵ^* is a shortened notation for the random phase of equation (6.3).

This initial value problem differs from the initial value problem solved in section 4 in that it is given as a function of x and y and its variation with time must be found and in that the crests are not perpendicular to the sides of the rectangle which bound the function.

Apart from modulation effects, the rectangular area under consideration can be moved off in the direction θ_{2q+1} with a group velocity determined by μ_{2m+1}. Its position at any time could be determined, and if it were present at a given time at a given point, then simple sine waves of that frequency traveling in that direction would be observed.

The raft at x_0,y_0 responds to the waves as a function of time. Suppose that the waves at the time t_{ob} are observed at the raft. Then the many envelopes of the terms in the partial sum will be spread out in different directions and positions with respect to their source, depending upon their group velocity, $g/2\mu_{2m+1}$, and the direction in which they are traveling, θ_{2q+1}.

Some of the envelopes will pass above the point of observation and some will pass below the point of observation. The energy associated with these waves can thus never cause any disturbance at the point

x_0,y_0. Other envelopes will have traveled so fast with their group velocity that their rear edge will already have passed the point of observation. Still others will have traveled so slowly that they will not yet have arrived at the point of observation.

16.3. A Wave Forecasting Filter

The frequency and direction of a given spectral component must both lie within a certain range of values if that sinusoidal wave is to be present at the point and time of observation. The bounds of the frequency are given by equation (16.2) and the bounds of the angle are given by equation (16.3)

$$\frac{gt_{ob}}{2(R_0 + F)} < \mu < \frac{gt_{ob}}{2R_0} \tag{16.2}$$

$$\tan^{-1}\left(\frac{y_0 - \frac{W_s}{2}}{x_0}\right) < \theta < \tan^{-1}\left(\frac{y_0 + \frac{W_s}{2}}{x_0}\right) \tag{16.3}$$

The lower bound of the frequency given in (16.2) is that frequency that can travel a distance equal to $R_0 + F$ in the time t_{ob}, and hence the rear edges of all envelopes associated with the frequencies less than this value will already have passed. An interpretation can be placed on each of the other bounds by a similar analysis. A more careful analysis would show that these values are not exactly right because of the rectangular shape of the envelope, but the errors made in simplifying the analysis tend to cancel.

As the mesh of the net in equation (6.3) is made small, more and more components fall within the bounds given by equations (16.2) and (16.3). In the limit, the spectrum of the swell observed at x_0,y_0 at the time t_{ob} is given by

$$[A(\mu,\theta)]^2_{\text{swell}} = \begin{cases} [A(\mu,\theta)]^2 & \text{if the values of } \mu \text{ and } \theta \text{ are within the bounds given by equations (16.2) and (16.3)} \\ 0 & \text{otherwise} \end{cases} \tag{16.4}$$

In electronics, such an operation on a spectrum is equivalent to passing the noise voltage associated with that spectrum through a band-pass filter. The spectrum of the output noise is then determined by the effect of the filter.

Hence, by analogy, equations (16.2) and (16.3) define a wave forecasting filter which determines the spectrum of the swell from the spectrum of the sea waves in the original storm. Other wave forecasting filters have been obtained for what correspond to other initial value problems. They all depend on locating the position in space and time of the various spectral components in the sea waves at the time that the strong wind ceases, and then propagating these components with an appropriate direction and speed.

16.4. Swell Characteristics

As x_0, y_0, and t_{ob} are varied by ± 10 kilometers and by ± 10 min, the filter characteristics do not vary very much. Consequently, the swell can be represented by a quasi-stationary Gaussian process with a spectrum given by equation (16.4). With large variations of x_0, y_0, and t_{ob}, the characteristics of the swell can be studied over the whole area into which the waves travel.

To simplify the discussion, let y_0 be zero and let x_0 be fairly large. Then, at a given time of observation, the frequency band width is given by equation (16.4) and the angular band width is approximately given by equation (16.5)

$$\Delta\mu = \frac{Fgt_{ob}}{2x_0(x_0 + F)} \tag{16.4}$$

$$\Delta\theta = \frac{W_s}{x_0} \tag{16.5}$$

If a given frequency is present at x_0 at t_{ob}, and if x_0 is doubled, then t_{ob} must be doubled for that frequency to be present at the new point. Doubling x_0 and t_{ob}, thus, nearly halves $\Delta\mu$ and $\Delta\theta$. The energy, which is the integral over the swell spectrum is decreased to one-fourth of its first value. Hence the height of the waves with a given swell "period" is cut in half by the effect of the filter if the distance from the generating area is doubled. The swell also becomes more regular and more long-crested.

At a point of observation, first high-period waves are observed and then medium-period waves arrive. Finally, the low-period waves arrive. The greater the distance from the point of observation, the longer it takes for the waves to pass. Since the low-frequency end of the swell spectrum has a sharp jump with increasing frequency, the swell arrives suddenly and then, as the higher frequencies arrive, the waves die down in height with increasing time.

The various filters which have been derived have been tested on actual weather situations by Pierson, Neumann, and James [50] and by James [64], and the results check well with available observations.

The spectra given by Barber and Ursell [61] confirm these results qualitatively but, since generating areas are irregular in shape and often bounded by areas of lower winds, the results are not as clear cut as the simple case described above would require. Correspondence with R. A. Wooding has also shown diagrams which illustrate the effect of such filters.

Data on the average values of fetch lengths and storm widths have recently been given by DeLeonibus.* Based on a study of five years of data over the North Atlantic Ocean for the winter months, DeLeonibus found that the average fetch length was 430 nautical miles and the average fetch width was 330 nautical miles for generating areas where the winds were Beaufort force seven or higher and which lasted for 24 hours or more.

Swell forecasts were prepared by DeLeonibus based on these average dimensions of the generating area with the use of a filter quite similar to the one defined by equation (16.4). The heights of the highest swell to pass given points in the decay area were found to agree quite well with the heights of the swell forecasted by applying the same assumed conditions to a forecast of significant height based on the methods of Sverdrup and Munk [56] and Bretschneider.†

16.5. The Period Increase of Ocean Swell

Since the energy in the band just above μ_i is the greatest in a partially developed sea, and since the energy is a maximum at $\mu_{\max}$ in a fully generated sea, the highest swell will be associated with these "periods." The average "period" in the generation area is considerably less than these by virtue of equation (9.2). Thus the "period" of the highest swell increases with increasing distance from the generation area until it equals either $2\pi/\mu_i$ or $2\pi/\mu_{\max}$, depending on the spectrum at the source, but it never exceeds this value because the spectral energy at lower frequencies is either absent or very low. In general, the "period" of the highest swell will not exceed 1.414 times the "period" of the waves in the generating area.

16.6. Friction Effects

The decrease of wave height with distance traveled from the source has mainly been explained solely on the basis of dissipation effects. An example of such an explanation is given by Groen and Dorrestein [65]. The effect of the filters given above appears to be adequate in describing the decrease in wave height without recourse to any added effect of

* DeLeonibus, P. S. (1955). Climatological data on the generation and propagation of waves in the North Atlantic. *in Proc. 1st Conf. Ships and Waves,* Council on Wave Research, The Engineering Foundation (in press).

† Bretschneider, C. L. (1952). Revised wave forecasting relationships. *Proc. 2nd Conf. Coastal Engineering,* Council on Wave Research, The Engineering Foundation.

friction. Dispersion and angular spreading must occur, so it must at least be conceded that the effect of friction has been overestimated by a large amount.

If a high wind opposes a traveling swell, the local waves generated by the wind oppose the traveling waves. Every time a breaker occurs, part of the energy dissipated by the turbulent motion may be taken from the swell system, thus decreasing its height.

The filters predict the eventual arrival of short-period waves. A wave with a period of 5 sec takes three times as long as a wave with a period of 15 sec to travel a given distance. Changing meteorological conditions may thus mask their arrival. New opposing winds may spring up to cut them down. In addition, pressure recorders may not even detect them. But if the winds are light, and if an attempt to detect these waves is made, the author believes that they will be found. Spectra observed by Barber and Ursell [61] cover periods from over 20 sec to less than 8 sec for waves known to have traveled over 500 nautical miles.

The effect of friction is certainly not well enough understood, however, to decide conclusively upon its magnitude. More study of this problem is needed.

16.7. Added Comments

These filters also explain the results obtained by Fuchs [30]. To a linear approximation, the short crested waves which he treats are simply the sum of two discrete sinusoidal waves of the same finite amplitude and of the same frequency. These two waves are traveling at plus and minus 45 degrees to the x-axis. Thus, when the stationary phase approximation is used to solve the initial value problem, the result is that the waves arrive at the sides and not directly in front of the disturbance.

17. Practical Methods for Observing and Forecasting Ocean Waves

17.1. The Synthesis

With an adequate working definition of what waves really are and with the theoretical spectra of Neumann [38], the statistical results of many workers, and the concept of the wave forecasting filter, it becomes possible to prepare a practical manual for forecasting ocean waves. The foundation of this manual is Neumann's theoretical spectra, for without them all of the above would just be theory with no hope of immediate practical application. Such a manual has been prepared by Pierson, Neumann, and James [50], and a revised version is soon to be published by the Hydrographic Office.

17.2. Contents

The manual just described presents the above theories in a simpler way for ready use and application. The subjects of wave generation, wave propagation, wave observation, ship motions, and wave refraction are treated, among others.

The curves given by equation (10.16) are plotted on an energy-frequency scale along with the curves given by equation (10.13). For wave generation problems, then, forecasts are prepared by simply reading off the appropriate energy values and computing the height distribution from a tabulation of data based on equation (9.1). The average "period" and the average "wavelength" are computed from equations (9.2) and (9.4).

For forecasting swell, four different filters are presented. The difference in the E values at two frequencies on the curves given by equation (10.16) is simply the energy between these two frequencies. Thus, when the two frequencies which are needed in each filter have been evaluated, the energy in the swell, neglecting the angular spreading effect, can be found. The angular spreading effect is found from the number given by evaluating equation (17.1) from diagrams which involve the appropriate angular measurements

$$K_A = \frac{2}{\pi} \int_{\tan^{-1}\left(\frac{y_0 - \frac{W_s}{2}}{X_0}\right)}^{\tan^{-1}\left(\frac{y_0 + \frac{W_s}{2}}{X_0}\right)} (\cos \theta)^2 d\theta \tag{17.1}$$

When the energy in the frequency band is multiplied by this factor, the full effect of the filter is accounted for, and the energy of the swell has been found. The frequency band determines the "period" band of the swell, and the E value determines the height forecast on the basis of equation (9.1).

17.3. Tests Needed

These forecasting methods have been tested and compared carefully with other methods. They sometimes give spectacular results, as in the case of a correct forecast of very rapid decreases of wave height in a generating area due to the effect of a filter which propagates the rear edge of each spectral component past the point of observation with its appropriate group velocity. The tests and comparisons have shown no major failures.

The methods still need further testing, and the theory needs to be critically reviewed by those familiar with the problem of wave forecasting. One reason for the preparation of this paper has been to present these theories as a unit for such study.

18. Unsolved Problems

18.1. Theory Not Complete

There are many problems in the theory of wind generated gravity waves which still need to be solved. A few of the more important ones will be described in what follows. The theory is not completely verified by observation, and it will eventually stand or fall or be modified on the basis of such observations.

18.2. A Partial List

The theoretical representation given in section 6 has not been proved to represent ocean waves by direct observation such that an n-dimensional Gaussian distribution is shown to be the property of a particular realization. A wave record as a function of time is Gaussian. The derivative as a function of space along a line is nearly Gaussian, and hence at an instant in time over the x,y-plane the waves have the correct properties. Ship motions are Gaussian, so the forcing seaway should be Gaussian on a line in the x,y,t-space. However, all of these observations are essentially marginal distributions of the more general case proved by the theorem. The above verification data are necessary conditions for the theorem to be true in nature, but they are not sufficient to prove by observation that the theorem is true in nature. However, since this representation of the waves explains so many things, it is difficult to see how the theorem can fail to represent actual waves apart from nonlinear effects.

Other statistical properties of the waves are unknown. The space-time statistical properties are not known. The probability distribution function of the times between zeros is not connected theoretically to the shape of the spectrum.

In wave generation theory, problems of variable winds both in time and space need to be treated. The high-frequency end of the spectrum appears to be limited in growth by continual breaking at the crests. Each time a wave breaks, Neumann believes that its forward momentum is added to the surface water. The currents which may be generated in this way need study. The generation of waves by moving fetches is being studied. More spectra of sea waves need to be obtained to verify directly (instead of indirectly) the theoretical spectra of Neumann.

In the theory of wave propagation, series of well documented spectra which illustrate the filters are needed. The angular variation of the spectra needs to be measured. Filters for moving generation areas need to be derived. The effect of friction needs to be carefully separated from the effect of dispersion and angular spreading and measured.

There are also a host of problems connected with applying these results to the practical design of ships (see Lewis [66]) and sea planes and with the utilization of these results in the design of offshore structures.

The results of this paper point the way toward the start of a solution to these problems since the complexity of actual wind generated gravity waves has at last yielded to analysis. Considerable progress has been made since that time when Lord Rayleigh said "The basic law of the seaway is the apparent lack of any law," and since the time just six years ago when Mason asked "What is an ocean surface wave?"

Acknowledgments

The results on which this paper is based could not have been obtained without the research conducted for the Beach Erosion Board, Corps of Engineers, U. S. Army. The opportunity to work on this subject was first given to the author by the Board, and their support since 1949 is very much appreciated. The Office of Naval Research has sponsored other aspects of this work in connection with the theory of ship motions and the theory of waves in deep water, and the Bureau of Aeronautics Project AROWA sponsored the preparation of the wave manual.

I would like to thank my colleague, Dr. Gerhard Neumann, for his suggestions and comments on this text and for the stimulating discussions we have had on this subject.

To formulate a problem is the first step in solving it. I would like to thank J. W. Tukey for asking the question "Is it Gaussian?" and Leo Tick for asking the question "What is the distribution of $\eta(x,y,t)$ at a set of arbitrarily chosen points?"

It is not possible to list the names of all the other people who have helped in the preparation of this paper, but their help is nevertheless appreciated.

List of Symbols

A amplitude of a simple harmonic progressive wave

A_N amplitude of cosine term in $J_N(t)$

B_N amplitude of sine term in $J_N(t)$ ($A_\infty = A$, $B_\infty = B$; A not same as above)

A_{mq} amplitude of term in sum of simple harmonic progressive waves

C speed of a simple harmonic progressive wave

C_1 a constant

C a constant equal to 3.05×10^4 cm^2 sec^{-5} (section 10)

C_g group velocity associated with a simple harmonic progressive wave

C_N amplitude of sinusoidal term in $J_N(t)$ ($C_\infty = C$, not same as above)

D_w the duration of a disturbance, in seconds

E the area (or volume) in cm^2 under the energy spectrum

E^* the expected value of

$\left.\begin{matrix}E_1\\E_2\end{matrix}\right\}$ constants determined from $E_1(\tau)$ and $E_2(\tau)$

E_F area under Neumann's spectrum from μ_i to ∞ when wave generation is controlled by the fetch

E_t area under Neumann's spectrum from μ_i to ∞ when wave generation is controlled by the duration

F length of the area of wave generation in the direction of the wind

H^* maximum of heights of waves with period $\tilde{T}$

$\bar{H}$ average wave height; the average of the crest to trough heights of all of the waves in a wave record

$\bar{H}_{1/3}$ significant wave height; the average of the crest to trough heights of the one-third highest waves in a wave record

$\bar{H}_{1/10}$ average of the crest to trough heights of the one-tenth highest waves in a wave record

I a number from the ensemble of values of a random integral

K constant

K_{SM} empirical constant determined by Sverdrup-Munk to fit v^2 law

K_N empirical constant determined by Neumann to fit $v^{2.5}$ law

K_A angular spreading factor

L wavelength of a simple harmonic progressive wave

$\tilde{L}$ distance along a line in the x,y-plane between two successive zero up crosses at an instant of time

$\bar{\tilde{L}}$ average of the values of $\tilde{L}$

$\bar{\tilde{L}}_{\max}$ average of the distances between two successive maxima measured along a line at an instant of time

L_h unsmoothed estimate of energy in a band of $[A(\mu)]^2$

M constant of integration

N constant of integration

N positive integer

Q_p lagged cross products of a wave record read at discrete points

$\Re e$ real part of a complex expression

R_0 distance from center of forward edge of generation area to point of observation

T period of a simple harmonic progressive wave

$\tilde{T}$ time interval between two successive zero up crosses in a wave record obtained at a fixed point as a function of time, or wave "period"

$\bar{\tilde{T}}$ average of the values of $\tilde{T}$; average "period"

$\bar{\tilde{T}}_{\max}$ average of the time intervals between successive maxima in a wave record obtained at a fixed point as a function of time

$T_{\max}$ period corresponding to the frequency at which the energy spectrum is a maximum

U_h smoothed estimate of the energy in a band of the energy spectrum

W_s width of an area of wave generation measured perpendicular to the wind

$[A(\mu)]^2$ energy spectrum of the waves when observed as a function of time at a fixed point [cm^2-sec]

$[A(\mu,\theta)]^2$ energy spectrum of the three-dimensional stationary Gaussian process which represents wind generated gravity waves in deep water [cm^2-sec/radian]

$[A(\mu,\theta)]^2_{\text{swell}}$ energy spectrum of the swell

$[A_R(\mu)]^2$ energy spectrum of the waves observed at a fixed point after refraction

$[A_{PR}(\mu)]^2$ energy spectrum of the pressure record of the waves observed by a pressure recorder on the bottom in water of depth h

$[A_R(\mu,\theta)]^2$ energy spectrum of the waves after refraction (qualified)

$[A(\mu_0;\theta_D)]^2$ energy spectrum of the waves as observed along a line at an angle of θ_D degrees to the wind direction

$E_1(\tau)$ a function which can be obtained from the energy spectrum

$E_2(\tau)$ a function which can be obtained from the energy spectrum

$F(x,y)$ a function which determines the properties of the envelope of the finite wave train in the y-direction as a function of x

$G(x,t)$ a function which determines the properties of the envelope of the finite wave train at various times as a function of x

$I(\mu^2h/g)$ Itcoth (μ^2h/g); the iterated hyperbolic cotangent of μ^2h/g

$J(t)$ ensemble of all possible sinusoidal variations in time with amplitudes chosen from the Rayleigh distribution and phases at zero time chosen uniformly between 0 and 2π

$\overline{\text{K.E.}}^t$ kinetic energy of the wave motion averaged over time at any fixed point

$\overline{\text{K.E.}}^{x'}$ kinetic energy of the wave motion averaged along a line at an instant of time

$[K(\mu,\theta)]^2$ spectrum amplification function

$P(-\infty < A < K)$ probability that A will be between $-\infty$ and K

$Q(p)$ representation for twice the covariance

$R^*(\eta(x_j,y_j,t_j),\ \eta(x_k,y_k,t_k))$ covariance function for a partial sum

$R(\eta(x_j,y_j,t_j),\ \eta(x_k,y_k,t_k))$ covariance function in the limit

$[S_x(\mu,\theta)]^2$ slope spectrum in the x-direction

$[S_y(\mu,\theta)]^2$ slope spectrum in the y-direction

a_1, a_2, a_3, a_4 constants

g acceleration of gravity

h depth of the water in section 3, subsection 13.4, and section 15.

h index of summation in equations (13.2) and (13.3)

i $\sqrt{-1}$ in complex numbers

i, j, k indices of summation

m, n, p, q indices of summation

r, s range of summation

t time

t duration of the wind in equation (10.13)

t' time (as variable of integration)

t^* translation in time

t_{ob} time of observation

v wind velocity

$x,y,z,$ rectangular Cartesian coordinate system, positive x often taken in the direction toward which the wind is blowing, z positive upward

x', y' variables of integration in section 4

x', y' rotated coordinates in section 6

x^*, y^* translation coordinates

x_0, y_0 coordinates of point of observation after waves have been refracted (section 15)

$x_0\ y_0$ coordinates of point of observation with respect to center of forward edge of generating area (section 17)

$a(\mu,\theta)$ cosine Fourier spectrum
$b(\mu,\theta)$ sine Fourier spectrum
$f(\mu)$ a function of μ defined in probability terms
$f(Z^i_{mq})$ probability distribution function of Z^i_{mq}
Z^i_{mq} a random variable
Δ a small increment
ΔE energy in $[A(\mu)]^2$ between $2\pi(n - \frac{1}{2})/\tau$ to $2\pi(n + \frac{1}{2})/\tau$
$\Gamma(\mu,\theta)$ Jacobian of $\Theta^*(\mu,\theta_D)$ and μ
$\Theta(\mu,\theta)$ the direction function
$\Theta^*(\mu,\theta_D)$ inverse of the direction function
α variable angle between 0 and 2π
α, β transformation variables and variables of integration
γ. δ variables of integration
ϵ phase of a simple harmonic progressive wave
$\epsilon(\mu)$ random phase over μ-axis
$\epsilon(\mu,\theta)$ random phase over μ,θ-plane
η departure of the free surface from the $z = 0$ plane
$\eta(t)$ free surface as a function of time at a fixed point
$\eta(x,y,t)$ free surface as a function of space and time
θ direction toward which wave is traveling when measured in a counterclockwise direction with respect to the positive x-axis.
θ_D angle through which x' and y' are rotated
θ_0 angle measured in counterclockwise direction with respect to positive x' axis.
θ_R angle toward which wave is traveling after refraction measured with respect to x_0 after refraction
μ (time) frequency in radians per second ($\mu = 2\pi/T$)
μ_i frequency of intersection
$\mu_{\max}$ frequency at which the energy spectrum is a maximum
$\bar{\bar{\mu}}$ one-half the average "frequency" of the zeros of a stationary Gaussian process as a function of time
$\mu_{S\max}$ frequency at which the slope spectrum is a maximum
ν (space) frequency in radians per centimeter. ($\nu = 2\pi/L$)
ν_0 (space) frequency in radians per centimeter when line along which distances are measured lies at an angle of θ_0 to the x'-axis
ξ variable of integration
ρ density of the water
σ variable of integration
σ_u^2 variance of the observed distribution of up-wind slopes (observed)
σ_c^2 variance of the observed distribution of cross-wind slopes (observed)
σ_x^2 variance of sea surface slopes in the x-direction (upwind theoretical)
σ_y^2 variance of the sea surface slopes in the y-direction (crosswind theoretical)
τ length of a sample wave record in seconds
ϕ potential function of a simple harmonic progressive wave
ψ phase of finite wave train
ω variable of integration

References

1. Mason, M. A. (1949). Ocean wave research and its engineering applications, *in* "Ocean Surface Waves." *Ann. N.Y. Acad. Sci.* **51,** 523–532.
2. Thorade, H. (1931). Probleme der Wasserwellen. *Prob kosm. Phys.* **12, 14,** 219 pp.
3. Deacon, G. E. R. (1949). Recent studies of waves and swell, *in* "Ocean Surface Waves." *Ann. N.Y. Acad. Sci.* **51,** 475–482.
4. Klebba, A. A. (1946). Progress report on shore based wave recorder and ocean wave analyzer. Memorandum for file. Woods Hole Oceanographic Institution.
5. Rudnick, P. (1951). Correlograms for Pacific Ocean waves. Proceedings of the 2nd Berkeley Symposium on Mathematical Statistics and Probability, pp. 627–638. University of California Press, Berkeley, Calif.
6. Birkhoff, G., and Kotik, J. (1952). Fourier analysis of wave trains, *in* "Gravity Waves." *Natl. Bur. Standards (U.S.) Circ.* **521,** 243–253.
7. Pierson, W. J., Jr. (1952). A unified mathematical theory for the analysis, propagation and refraction of storm generated ocean surface waves. Parts I and II. Research Division, College of Engineering, New York University, Department of Meteorology and Oceanography. Prepared for Beach Erosion Board, Department of the Army, and Office of Naval Research, Department of the Navy.
8. St. Denis, M., and Pierson, W. J., Jr. (1953). On the motions of ships in confused seas. Transactions of the Society of Naval Architects and Marine Engineers. Vol. 61, pp. 280–357.
9. Proudmann, J. (1953). Dynamical Oceanography. John Wiley and Sons, New York.
10. Biesel, F. (1952). Study of wave propagation in water of gradually varying depth, *in* "Gravity Waves." *Natl. Bur. Standards (U.S.) Circ.* **521,** 221–234.
11. Lamb, H. (1932). Hydrodynamics, 6th ed. Dover Publications, New York.
12. Stokes, C. G. (1847). On the theory of oscillatory waves. *Trans. Cambridge Phil. Soc.* **8,** 441.
13. Davies, T. V. (1951). The theory of symmetrical gravity waves of finite amplitude. *Proc. Roy. Soc. (London)* **A208,** 1951.
14. Davies, T. V. (1952). Symmetrical finite amplitude gravity waves, *in* "Gravity Waves." *Natl. Bur. Standards (U.S.) Circ.* **521,** 55–60.
15. Airy, G. B. (1824). Tides and waves. *Encyclopaedia Metropol. (London)* p. 241.
16. Merian, J. R. (1828). Über die Bewegung tropfbarer Flüssigkeiten in Gefässen. Basel.
17. Green, G. (1839). Note on the motions of waves in canals. *Trans. Cambridge Phil. Soc.* **7,** 87.
18. Lowell, S. C. (1949). The propagation of waves in shallow water. *Communs. Pure Appl. Math.* **2,** No. 2–3, 275–291 (abridged).
19. Lacombe, H. (1952). The diffraction of a swell. A practical approximate solution and its justification, *in* "Gravity Waves." *Natl. Bur. Standards Circ.* **521,** 129–140.
20. Johnson, J. W., and Blue, F. L. (1949). Diffraction of water waves passing through a breakwater gap. *Trans. Am. Geophys. Un.* **30,** No. 5.
21. Keller, J. (1949). The solitary wave and periodic wave in shallow water, *in* "Ocean Surface Waves." *Ann. N.Y. Acad. Sci.* **51,** 345–350.
22. Friedrichs, K. O. (1948). Water waves on a shallow sloping beach. *Communs. Pure Appl. Math.* **1,** No. 2, 109–134.
23. Peters, A. S. (1952). Solution of a mixed boundary value problem for

$$\nabla^2\phi - K^2\phi = 0$$

in a sector. *Communs. Pure Appl. Math.* **5,** 87–108.

24. Eckart, C. (1952). The propagation of gravity waves from deep to shallow water, *in* Gravity Waves. *Natl. Bur. Standards (U.S.) Circ.* **521,** 165–174.
25. Johnson, J. W., O'Brien, M. P., and Isaacs, J. D. (1948). Graphical construction of wave refraction diagrams. Hydrographic Office (U.S. Navy), Publication No. 605.
26. Pierson, W. J., Jr. (1951). The interpretation of crossed orthogonals in wave refraction phenomena. Technical Memorandum No. 21, Beach Erosion Board, Washington, D. C.
27. Arthur, R. S., Munk, W. H., and Isaacs, J. D. (1952). The direct construction of wave rays. *Trans. Am. Geophys. Un.* **33,** No. 6.
28. Luneberg, R. M. (1944). Mathematical theory of optics. Brown University, Providence, R. I. (Mimeographed lecture notes prepared for a course entitled "Advance instruction and research in mechanics." No longer available.)
29. Coulson, C. A. (1943). Waves. Interscience Publishers, New York.
30. Fuchs, R. A. (1952). On the theory of short crested oscillatory waves, *in* "Gravity Waves." *Natl. Bur. Standards (U.S.) Circ.* **521,** 165–174.
31. Pierson, W. J., Jr. (1952). On the propagation of waves from a model fetch at sea, *in* "Gravity Waves." *Natl. Bur. Standards (U.S.) Circ.* **521,** 175–186.
32. Lévy, P. (1948). Processes Stochastiques et Mouvement Brownien. Gauthier-Villars, Paris (suivi d'une note de M. Loeve). (See Chapter 4.)
33. Rice, S. O. (1944). Mathematical analysis of random noise. *Bell System Tech. J.* **23,** 282–332; (1945) **24,** 46–156.
34. Tukey, J. W. (1949). The sampling theory of power spectrum estimates. Symposium on Applications of Autocorrelation Analysis to Physical Problems, Woods Hole, Mass., June 13–14. (Office of Naval Research, Washington, D.C.)
35. Wiener, N. (1949). Extrapolation, Interpolation and Smoothing of Stationary Time Series. John Wiley and Sons, New York.
36. Neumann, G. (1952). Über die komplexe Natur des Seeganges. I. Neue Seegangsbeobachtungen im Nordatlantischen Ozean, in der Karibschen See und im Golf von Mexico (M.S. Heidberg, Oct. 1950–Feb. 1951). *Deut. Hydrog. Z.* **5,** 95–110; II. Das Anwachsen der Wellen unter dem Einfluss des Windes. *Deut. Hydrog. Z.* **5,** 252–277.
37. Neumann, G. (1952). On wind generated ocean waves with special reference to the problem of wave forecasting. Research Division, College of Engineering, New York University, Department of Meteorology and Oceanography. Prepared for the Office of Naval Research.
38. Neumann, G. (1953). On ocean wave spectra and a new method of forecasting wind-generated sea. Technical Memorandum No. 43, Beach Erosion Board, Washington, D. C. (Corps of Engineers, Department of the Army).
39. Neumann, G. (1954). Zur Charakteristik des Seeganges. *Arch. Meteorol. Geophys. Biokl.* **A7,** 352–377.
40. Courant, R. (1937). Differential and Integral Calculus, Vols. I and II. Interscience Publishers, New York.
41. Jahnke, E., and Emde, F. (1945). Funktionentafeln mit Formeln und Kurven, 4th rev. ed. Dover Publications, New York.
42. Pierson, W. J., Jr. (1954). The representation of ocean surface waves by a three dimensional stationary Gaussian process. (Manuscript prepared for the Office of Naval Research.)

43. Cramer, H. (1937). Random Variables and Probability Distributions, Tract No. 36. Cambridge University Press, England. (See Chapter 10.)
44. Marks, W. (1954). The use of a filter to sort out directions in a short crested Gaussian sea surface. *Trans. Am. Geophys. Un.* **35,** No. 5.
45. Pierson, W. J., Jr., and Marks, W. (1952). The power spectrum analysis of ocean wave records. *Trans. Am. Geophys. Un.* **33,** No. 6.
46. Pierson, W. J., Jr. (1954). An interpretation of the observable properties of sea waves in terms of the energy spectrum of the Gaussian record. *Trans. Am. Geophys. Un.* **35,** No. 5.
47. Pierson, W. J., Jr. (1954). Visual Wave Observations. (Unpublished manuscript.)
48. Barber, N. F. (1950). Ocean waves and swell. Lecture published by *Inst. of Elec. Engrs. (London)* 22 pp.
49. Longuet-Higgins, M. S. (1952). On the statistical distribution of the heights of sea waves. *J. Mar. Res.* **11,** No. 3, 245–266.
50. Pierson, W. J., Jr., Neumann, G., and James, R. W. (1953). Practical methods for observing and forecasting ocean waves by means of wave spectra and statistics. Prepared for Bureau of Aeronautics Project AROWA, Contract No. N189s-86743. (Revised version to be published by the Hydrographic Office.)
51. Watters, J. K. A. (1953). Distribution of the heights of ocean waves. *New Zealand J. Sci. Technol.* **B34,** No. 5.
52. Putz, R. R. (1954). Statistical analysis of ocean waves. *Inst. Eng. Res. Univ. Calif.* Ser. **3,** No. 350.
53. Putz, R. R. (1954). Statistical analysis of wave records. *Inst. Eng. Res. Univ. Calif.* Ser. **3,** No. 359.
54. Seiwell, R. H. (1948). Results of research on surface waves of the Western North Atlantic, II. *Papers Phys. Oceanog. Meteorol. Mass. Inst. Technol. and Woods Hole Oceanog. Inst.* **10,** No. 4, 30–51.
55. Weigel, R. L. (1949). An analysis of data from wave recorders on the Pacific coast of the U.S. *Trans. Am. Geophys. Un.* **30,** 700–704.
56. Sverdrup, H. U., and Munk, W. H. (1947). Wind, sea and swell: Theory of relations for forecasting. Hydrographic Office (U.S. Navy), Publication No. 601.
57. Cornish, V. (1935). Ocean Waves and Kindred Geophysical Phenomena. Cambridge University Press, England.
58. Arthur, R. S. (1949). Variability in direction of wave travel, *in* "Ocean Surface Waves." *Ann. N.Y. Acad. Sci.* **51,** 511–522.
59. Dearduff, R. F. (1953). A comparison of observed and hindcast wave characteristics off southern New England. *Bull. Beach Eros. Bd. Wash.* **7,** No. 4.
60. Cox, C., and Munk, W. H. (1952–1953). The measurement of the roughness of the sea surface from photographs of the sun's glitter, Parts I, II, and III. AF Technical Reports Nos. 1, 2, and 3. S.I.O. References 52-61, 53-53 and 54-10.
61. Barber, N. F., and Ursell, F. (1948). The generation and propagation of ocean waves and swell. I. Wave periods and velocities. *Phil. Trans. Roy. Soc. (London)* **240,** No. 824, 527–560.
62. Stoker, J. J., and Peters, A. S. (1954). The Motion of Ship, as a Floating Rigid Body, in a Seaway. The New York University Institute of Mathematical Sciences. IMM-NYU 203.
63. Pierson, W. J., Jr., Tuttell, J. J., and Woolley, J. A. (1953). The theory of the refraction of a short crested Gaussian sea surface with application to the northern New Jersey coast. *Proc. 3rd Conf. on Coastal Engineering* (J. W. Johnson, ed.), 86–108.

64. James, R. W. (1954). An example of a wave forecast based on energy spectrum methods. *Trans. Am. Geophys. Un.* **35,** No. 1.
65. Groen, P., and Dorrestein, R. (1950). Ocean swell, its decay and period increase. *Nature* **165,** 445–447.
66. Lewis, E. V. (1955). Ship model motions in regular and irregular seas. Rept. No. 531, Experimental Towing Tank, Stevens Inst. of Technology (Preliminary copy).

Geological Chronometry by Radioactive Methods*

J. LAURENCE KULP

Lamont Geological Observatory (Columbia University), Palisades, N. Y.

1. Introduction

The large increase in our knowledge of nuclear physics and radiochemistry occasioned by the Manhattan Project during World War II has had important ramifications in most other fields of experimental science. Not only did the Project produce many new ideas and techniques, but also numerous young scientists who were interested in the application of these techniques to other areas of study. In earth science, geochemistry

* Lamont Geological Observatory Contribution No. 125.

and nuclear geophysics have been profoundly affected by these developments. This is particularly true of the general area of age determination of rocks and minerals. More research in quantitative geochronometry has been conducted during the past decade than in the previous five. Since the time dimension is of central importance to geological theory, the accomplishments attending these researches will produce extensive advances in the whole field of earth science. It appears timely, therefore, to review the present status of knowledge in the absolute measurement of geologic time.

Although relative estimates of geological time can be based on such phenomena as rates of sedimentation, the only absolute methods are based on radioactivity. The general principle of all isotopic clocks is simple. A radioactive isotope of element A, A*, decays to a stable isotope of element B, B*, at a known rate. This rate is independent of physical and chemical conditions that could have existed in the earth since its formation. If element A forms a mineral under certain geological conditions which is poor in element B, the ratio of radiogenic B* to the parent isotope A* will give the time since formation. An example would be the decay of Rb^{87} to Sr^{87}. Another way of utilizing the radioactive decay is to measure the specific activity of A* as a function of time, providing the activity at $t = 0$ is known. An example of this application is the Carbon-14 method. Still another method is to examine the ratio of B* to some other nonradiogenic isotope of B. In this case the ratio of A to B in a homogeneous reservoir must be known (e.g., the common lead method). Finally, the radioactive decay may be measured by some indirect effect such as the disorganization of a crystal lattice (metamict method).

The history of quantitative age determination is an instructive case history in the progress of science. Rutherford [1] first suggested that radioactivity could be used to determine geological time by the ratio of helium to uranium in a primary mineral. Within a year Boltwood [2, 3] proposed the ratio of lead to uranium as a more suitable index. It soon appeared that in highly radioactive minerals a large fraction of the helium is lost, and hence for the next thirty years the main effort in age determination was placed on the chemical analysis of radioactive minerals for uranium, thorium, and lead. Attempts were made to detect common lead contamination by tedious atomic weight determinations of the lead extracted from radioactive minerals, but this was largely ineffective. Some studies on the density of pleochroic halos in biotite and other minerals during this period proved only of academic interest. Thus in 1931 when the National Research Council issued the *Age of the Earth* volume [4], there were several hundred "chemical" ages (uranium–lead and thorium–lead), all of which are subject to question in terms of later developments. This pioneer work, however, established the fact that the earth was at least two billion years old and that an absolute geological time scale could be constructed.

In the 1930's the chemical age work continued at a reduced pace. The discovery of isotopes and the measurement of the isotopes of lead by Aston were the necessary preliminaries to the classic work of Nier [5, 6], at the end of the decade, in which he determined the isotopic composition of a number of radiogenic leads, and thus for the first time was able to correct quantitatively for common lead contamination and to utilize the three decay schemes of uranium and thorium to intercheck apparent ages. Also during this period a renewed effort was made to use the helium accumulation in rocks and minerals of intermediate to low uranium content as an age measuring device by Urry [7–10], Keevil [11, 12], and later by Keevil [13–17] and Hurley and Goodman [18, 19]. In retrospect too much time was spent "dating" samples and not enough on principles (Evans *et. al* [20]). The later work of Hurley pointed up the problems which remain to be solved before the helium clock can be considered reliable. The last item of importance in this decade was the discovery of the ionium method by Piggot and Urry [21], which is potentially applicable to deep sea sediments.

After the hiatus of World War II a greatly intensified effort in geochronometry occurred. Laboratories at M.I.T. and the University of Toronto which were active before the war expanded their programs. Several new laboratories placed a large emphasis on research in this field, including the Institute for Nuclear Studies at the University of Chicago, the U.S. Geological Survey, Geology Department at California Institute of Technology, and the Lamont Geological Observatory at Columbia University. Significant research in this area on a somewhat smaller scale has been initiated at Princeton, University of California, Yale, Pennsylvania State University, University of Pennsylvania, University of Wisconsin, and a number of the oil company laboratories. The net result of this increase in effort has been the discovery of a number of new methods, the study of the mechanics and assumptions underlying the methods, and the accumulation of new age determinations. Anomalies have also developed which give impetus to further work in the field. Although major advances have been made in geochronometry, it is far from a completed discipline. Table I lists most of the known clocks with their potential area of application.

In this chapter an attempt will be made to review the status of the significant isotopic clocks. There has been no attempt to cover the literature exhaustively (this is done annually in the Report of the Committee

TABLE I. Methods of age determination.

Name	Type of Material	Range (yrs)
A. Established methods—numerous reliable dates		
U, Th-Pb	Minerals containing at least 0.1% U or Th (Pegmatites, pitchblende veins, possibly granites)	1×10^6–T_0*
C^{14}	Carbon-bearing substance once in contact with photosynthetic cycle (wood, shell, peat, etc.)	0–45,000
B. Promising methods—under intensive study		
Rb-Sr	Rubidium-rich, strontium-poor minerals (Pegmatites-lepidolite, amazonite, Granites-biotite)	1×10^8–T_0
K-A	Most igneous and metamorphic rocks (Potassium feldspar, micas)	1×10^7–T_0
K-Ca	Certain high-K, low-Ca minerals (Lepidolite)	1×10^8–T_0
He	Igneous rocks (Mafic extrusives appear best)	1×10^6–T_0
Ionium	Deep sea sediments (Homogeneous normal deposits)	0–4×10^5
C. Other potential methods		
Metamict	Radioactive minerals	Variable
Thermoluminescence	Slightly radioactive minerals	?
Common lead	Galena	1×10^9–T_0
Tritium	Ground water	0–100
Spontaneous fission	Radioactive minerals	?
Sm	Rare earth minerals—pegmatites	?
Re^{187}	Molybdenite	?

* T_0 is the time of the origin of planet.

on Geologic Time [22] currently edited by the Chairman, J. P. Marble). The discussion will be concerned primarily with the developments of the past decade ending with the summer of 1954. The space allotted to each method is roughly proportional to the results achieved and the potential value to geology.

2. The Lead Methods

Until 1939, the "lead method" designated the age measurement based on the ratio of total lead to uranium plus thorium in a mineral. More recently this has been called the "chemical," "crude," or "uncorrected" age. With the discovery and measurement of the isotopic composition of lead, uranium, and thorium, from radioactive minerals several ratios became available which could be used as internal checks. The "lead method" now generally supplies the age interpreted from these isotopic ratios. Recently the change with time in the isotopic composition of common lead has been used as an index of age. This has been designated the "common lead" method. The "chemical" method is largely of historical interest because it does not provide any means of determining the influence of the factors which can alter the apparent age. It has been revived recently in a more elegant form by Larsen and co-workers [23] in an attempt to date zircons by determining the uranium and thorium by alpha counting the lead spectroscopically and by assuming the absence of common lead. This work extends the range of the chemical method to smaller concentrations of lead, but it does not avoid any of the inherent uncertainties.

The chemical lead method produced the dates which were used to calibrate the geologic time scale until the work of Nier [5, 6]. From then until the present the basic calibration has been derived from isotopic ratios involving uranium, thorium, and lead. As the primary method of age determination of older rocks it has been subjected to considerable investigation, which is continuing at this writing.

2.1. Principles

There are three ways that radiogenic lead may be formed: $U^{238} \rightarrow Pb^{206}$, $U^{235} \rightarrow Pb^{207}$, and $Th^{232} \rightarrow Pb^{208}$. The intermediate products and their respective half-lives are given in Tables II, III, and IV. In less than a million years after the deposition of a radioactive mineral, the three decay series are essentially in equilibrium and the amount of lead found bears a simple relationship to the age of the mineral. The age is then determined by the ratio of the stable end-product lead to any one of the members of the series which is in secular equilibrium. Thus the age may be determined equally well in principle by the ratios Pb^{206}/U^{238}, Pb^{206}/Ra^{226}, or Pb^{206}/Pb^{210}. Since chemical changes (such as leaching in the late history of a mineral) will affect these ratios in differing degrees, the measurement of several ratios permits quantitative evaluation of the factors which produce erroneous ages. In addition to the ratios of a single series, it is also possible to make use of the ratio Pb^{207}/Pb^{206}, since the ratio of U^{235}/U^{238} is geographically constant but the half-lives differ. These clocks involve different chemical characteristics, half-lives, and concentrations, so that agreement between them is strong evidence of the reliability of the calculated age.

The equations governing the change in these experimentally measurable ratios with time are straightforward and have been derived by Keevil [24]. The final equations modified for the latest constants are given in Table V. Convenient charts for reading the age directly, once given the isotopic ratios, have been published recently by Kulp *et al.* [25].

TABLE II. The uranium (U^{238}) series.

Isotope	Particle emitted	Particle energy (Mev.)	Half-life	
$_{92}U^{238}$(UI)	α	4.18	4.49 ± .01 × 10^9	Years
$_{90}Th^{234}$(UX_1)	β	0.205	24.101 ± .025	Days
		0.111		
$_{91}Pa^{234}$(UX_2)	β	2.32, 1.50	1.175 ± .003	Minutes
		0.60		
$_{92}U^{234}$(UII)	α	4.763	2.475 ± .016 × 10^5	Years
$_{90}Th^{230}$(Io)	α	4.68, 4.61	8.0 ± .3 × 10^4	Years
$_{88}Ra^{226}$(Ra)	α	4.777	1622 ± 1	Years
$_{86}Rn^{222}$(Rn)	α	5.486	3.825 ± .005	Days
$_{84}Po^{218}$(RaA)	α	5.998	3.050 ± .009	Minutes
$_{82}Pb^{214}$(RaB)	β	0.65	26.8 ± .1	Minutes
$_{83}Bi^{214}$(RaC)	α 0.04	5.46	19.72 ± .04	Minutes
	β 99.96	1.65, 3.17		
$_{84}Po^{214}$(RaC′)	α	7.680	163.7 ± .2	Microseconds
$_{81}Th^{210}$(RaC″)	β	1.8	1.32 ± .01	Minutes
$_{82}Pb^{210}$(RaD)	β	0.018	22.5 ± .4	Years
$_{83}Bi^{210}$(RaE)	β	1.17	4.989 ± .013	Days
$_{84}Po^{210}$(RaF)	α	5.298	138.374 ± .032	Days
$_{82}Pb^{206}$(RaG)			Stable	

TABLE III. The actinium (U^{235}) series.

Isotope	Particle emitted	Particle energy (Mev.)	Half-life	
$_{92}U^{235}$(AcU)	α	4.40, 4.58	7.13 ± .16 × 10^8	Years
		4.20, 4.47		
$_{90}Th^{231}$(Uy)	β	0.094, .302	25.6 ± .1	Hours
		0.216		
$_{91}Pa^{231}$(Pa)	α	5.00	3.43 ± .03 × 10^4	Years
$_{89}Ac^{227}$(Ac)	β	0.04	22.0 ± .3	Years
$_{90}Th^{227}$(RdAc)	α	6.00	18.6 ± .1	Hours
$_{88}Ra^{223}$(AcX)	α	5.70	11.2 ± .2	Days
$_{86}Rn^{219}$(An)	α	6.82	3.917 ± .015	Seconds
$_{84}Po^{215}$(AcA)	α	7.37	1.83 ± .04 × 10^{-3}	Seconds
$_{82}Pb^{211}$(AcB)	β	1.39, .50	36.1 ± 2	Minutes
	α	6.62		
$_{83}Bi^{211}$(AcC)	β	—	2.16 ± .03	Minutes
$_{84}Po^{211}$(AcC′)	α	7.43	0.52 ± .02	Seconds
$_{81}Tl^{207}$(AcC″)	β	1.44	4.79 ± .02	Minutes
$_{82}Pb^{207}$(Pb)			Stable	

TABLE IV. The thorium series.

Isotope	Particle emitted	Particle energy (Mev.)	Half-life	
$_{90}Th^{232}$(Th)	α	3.98	$1.39 \pm .02 \times 10^{10}$	Years
$_{88}Ra^{228}$(MsTh$_1$)	β	0.012	$6.7 \pm .1$	Years
$_{89}Ac^{228}$(MsTh$_2$)	β	1.15	$6.13 \pm .03$	Hours
$_{90}Th^{228}$(RdTh)	α	5.42	$1.90 \pm .01$	Years
$_{88}Ra^{228}$(ThX)	α	5.68	$3.64 \pm .01$	Days
$_{86}Rn^{220}$(Tn)	α	6.28	$54.53 \pm .04$	Seconds
$_{84}Po^{216}$(ThA)	α	6.77	$0.158 \pm .008$	Seconds
$_{82}Pb^{212}$(ThB)	β	0.355	$10.67 \pm .05$	Hours
$_{83}Bi^{212}$(ThC)	α	6.05	$60.48 \pm .04$	Minutes
	β	2.25		
$_{84}Po^{212}$(ThC′)	α	8.78	$0.29 \pm .01$	Microseconds
$_{81}Tl^{208}$(ThC″)	β	1.79	$3.1 \pm .1$	Minutes
$_{82}Pb^{208}$(ThD)			Stable	

TABLE V. Equation and constants used in lead method.

$$\frac{N_{207}}{N_{206}} = \frac{N_{235}}{N_{238}}\left[\frac{e^{\lambda_{235}T} - 1}{e^{\lambda_{238}T} - 1}\right] = \frac{1}{137.7}\left[\frac{e^{0.9722\times10^{-9}T} - 1}{e^{0.1541\times10^{-9}T} - 1}\right]$$

$$\frac{N_{206}}{N^\circ_{238}} = 1 - e^{-\lambda_{238}T} = 1 - e^{-0.1541\times10^{-9}T}$$

$$\frac{N_{207}}{N^\circ_{235}} = 1 - e^{-\lambda_{235}T} = 1 - e^{-0.9722\times10^{-9}T}$$

$$\frac{N_{208}}{N^\circ_{232}} = 1 - e^{-\lambda_{232}T} = 1 - e^{-0.4987\times10^{-10}T}$$

$T_{1/2}(U^{238}) = 4.51 \pm .01 \times 10^9$ yrs

$T_{1/2}(U^{235}) = 7.13 \pm .16 \times 10^8$ yrs

$\frac{N_{238}}{N_{235}} = 137.7 \pm .5$

$T_{1/2}(Th^{232}) = 1.39 \pm .02 \times 10^{10}$ yrs

N°_{238}, N°_{235}—number of atoms at $T = 0$

N_{238}, N_{235}, etc.—number of atoms at present

The problem of determining the concentration of common lead in a radioactive mineral would be hopeless, were it not for the presence of Pb^{204} which occurs in all common lead, but is not known to be radiogenic. By knowing the ratios of Pb^{204} to Pb^{206}, Pb^{207}, and Pb^{208} for the common lead of an area in which a radioactive mineral occurs, the amount of common lead of each isotope present in the mineral can be determined.

TABLE VI. Published lead-isotopic ages on radioactive minerals.

Sample	Locality	Common lead‡	Lead isotope ratios—Ages (million years)			
			206/238	*207/235*	*207/206*	*208/232*
N-1	Joachimsthal (Pitchblende)	*	244 ± 5	249 ± 22	242 ± 200	—
N-2	Katanga, Congo (Pitchblende)	*C*-2	610 ± 4	615 ± 6	626 ± 14	—
N-6	Katanga, Congo (Pitchblende)	*C*-2	567 ± 10	577 ± 16	620 ± 26	—
N-9	Beaver Lodge, N.W.T. (Pitchblende)	*C*-15	337 ± 2	389 ± 2	705 ± 15	—
N-10	Great Bear Lake (Pitchblende)	*C*-15	1220 ± 10	1282 ± 15	1400 ± 20	—
N-11	Bedford, N.Y. (Uranoan Zircon)	*L*-10	354 ± 4	355 ± 7	385 ± 43	—
N-12	Bedford, N.Y. (Uranoan Zircon)	*L*-10	329 ± 3	325 ± 8	308 ± 64	—
N-13	Besner, Ont. (Uraninite)	*C*-15	755 ± 4	780 ± 5	826 ± 15	812 ± 10
N-14	Huron Claim, Manitoba (Uraninite)	*C*-15	1555 ± 11	1990 ± 20	2490 ± 30	1090 ± 30
N-28§	Huron Claim, Manitoba (Monazite)	*C*-15	3180 ± 40	2840 ± 15	2610 ± 20	1830 ± 10
N-15	Wilberforce, Ont. (Uraninite)	*C*-15	1056 ± 10	1055 ± 15	1041 ± 20	1000 ± 20
N-17	Morogoro (Uraninite)	*C*-15	790 ± 6	743 ± 6	612 ± 29	—
N-18§	Aust, Agder, Norway (Cleveite)	*C*-15	1070 ± 5	1077 ± 6	1098 ± 20	855 ± 8
N-20	Brevig, Norway (Thorite)	*C*-15	—	—	—	238 ± 7
N-19	Pied des Monts, Quebec (Cleveite)	*C*-15	864 ± 5	875 ± 5	890 ± 30	—
N-21	Gullhögen, Sweden (Kolm)	*C*-15	378 ± 12	440 ± 105	800 ± 20	—
N-22	Parry Sound (Uraninite)	*C*-15	970 ± 5	1015 ± 10	1050 ± 30	955 ± 20
N-23	Glastonbury, Conn. (Samarskite)	*L*-10	253 ± 3	255 ± 4	293 ± 20	275 ± 4

TABLE VI. Published lead-isotopic ages on radioactive minerals. (*Continued*)

Sample	Locality	Common lead‡	Lead isotope ratios—Ages (million years)			
			206/238	*207/235*	*207/206*	*208/232*
N-24	Woods Mine, Colo. (Pitchblende)	‖	56 ± 5	60 ± 11	192 ± 300	—
N-25	Gilpin Co., Colo. (Pitchblende)	‖	58 ± 5	64 ± 10	293 ± 350	—
N-26§	Parry Sound (Thucolite)	*C*-15	269 ± 4	282 ± 7	433 ± 200	248 ± 9
N-27	Mount Isa, Australia (Monazite)	*C*-15	—	—	—	933 ± 12
N-29§	Las Vegas, Nev. (Monazite)	†	1723 ± 40	1547 ± 60	1300 ± 135	770 ± 10
*S-GS*63	Shinarump, No. 1 Claim, Grand Co., Utah (Pitchblende)	†	75 ± 5	—	—	—
*S-GS*64	Happy Jack Mine, San Juan Co., Utah (Pitchblende)	†	65 ± 5	—	—	—
S	Wood Mine, Gilpin Co., Colo. (Pitchblende)	—	66 ± 5	—	—	—
S	Iron Mine, Gilpin Co., Colo. (Pitchblende)	—	70 ± 5	—	—	—
K-1	Sunshine Mine, Idaho (Pitchblende)	—	710 ± 10	750 ± 10	850 ± 50	—
M-1	Bisundavi, Rajputana, India (Uraninite)	—	733 ± 10	733 ± 15	740 ± 30	940 ± 200
M-2	Soniana, Mewar State, India (Monazite)	—	660 ± 10	681 ± 20	865 ± 30	613 ± 6
H-1	Ebonite Tantalum Claims, Bikita District, So. Rhodesia (Monazite from pegmatite)	2600 million years lead (theo)	2675 ± 30	2680 ± 70	2680 ± 30	2645 ± 30

TABLE VI. Published lead-isotopic ages on radioactive minerals. (*Continued*)

Sample	Locality	Common lead‡	Lead isotope ratios—Ages (million years)			
			206/238	*207/235*	*207/206*	*208/232*
H-2	Jack Tin Claims, No. of Salsbury, So. Rhodesia (Monazite from pegmatite)	2600 million years lead (theo)	2260 ± 30	2470 ± 50	2650 ± 25	1940 ± 20
H-3	Irumi Hills, No. Rhodesia (Monazite concentrate from alluvium from "Lower Basement" rocks)	2600 million years lead (theo)	2040 ± 60	2330 ± 100	2620 ± 25	1390 ± 20
H-2	Soafia, Madagascar (Thorianite)	—	524 ± 5	493 ± 3	450 ± 10	465 ± 5
H-3	Betroka, Madagascar (Thorianite)	—	510 ± 5	522 ± 5	490 ± 20	437 ± 5
H-5	Bemasoandro, Madagascar (Uraninite)	—	390 ± 5	407 ± 7	485 ± 20	430 ± 5
C-16	Fish Hook Bay, Goldfields, Sask. (Pitchblende)	—	1485	1648	1850 ± 50	—
C-30	Martin Lake, Goldfields, Sask. (Pitchblende)	—	400	630	1580 ± 60	—
C-51	YY Concession, Goldfields, Sask. (Pitchblende)	—	190	260	940 ± 30	—
C-52	Gil group, Goldfields, Sask. (Pitchblende)	—	822	850	930 ± 20	—
C-53	50-AA-14, Goldfields, Sask. (Pitchblende)	—	592	664	920 ± 50	—
C-58	Bolger, Goldfields, Sask. (Pitchblende)	—	444	437	400 ± 50	
C-77	Pit No. 4, Wilberforce, Ont. (Uraninite)	—	1150	1110	1032 ± 10	1130

TABLE VI. Published lead-isotopic ages on radioactive minerals. (*Continued*)

Sample	Locality	Common lead‡	Lead isotope ratios—Ages (million years)			
			206/238	*207/235*	*207/206*	*208/232*
C-79	Calabogie, Ont. (Euxenite)	—	642	732	1010 ± 70	337
C-89	Witwatersrand, So. Africa (Uraninite)	—	1340	1650	2070 ± 50	—

* Galena, Saxony (Nier, *J. Am. Chem. Soc.* **60,** 1571, 1938).
† Vanadinite (Nier, *J. Am. Chem. Soc.* **60,** 1571, 1938).
‡ Common lead used in correction—from Table VIII unless otherwise noted.
§ Low either in uranium or thorium.
¶ Average galena, Colorado Plateau (personal communication, L. R. Stieff and T. W. Stern, U.S. Geol. Survey).
N—Nier [5, 6]
K—Kerr and Kulp [26]
M—Nier, National Research Council [27]
S—Stieff and Stern [28]
C—Collins, Farquhar, and Russell [29]
H—Holmes [30, 31]

2.2. Published Results

Primary uranium minerals for which isotopic and chemical analyses have been published are listed in Table VI [26–31]. In compiling this table the data were recomputed using the equation and constants given in Table V. The probable errors were estimated from the mass spectrometer data, the uncertainties in the half-lives and isotopic abundances of the uranium isotopes, and the uncertainty in the isotopic composition of the common lead used in making that correction. It does not include possible errors in the U^{235} half-life.

Possibly the most striking thing about these data is the lack of internal agreement between the various isotopic ages for the same sample. It is clear that additional factors are involved. One observation of particular importance is that the *207/235* ages are generally a little higher than the *206/238* ages, whereas the *207/206* ages are much higher. The explanation for this will be considered after the errors which affect the age measurements have been discussed.

2.3. Radon Leakage

Radon (Rn^{222}) which lies in the U^{238}–Pb^{206} decay chain has a half-life of 3.82 days. Therefore, it may migrate from the site of formation in a radioactive mineral. The radon that escapes will decrease the ultimate *206* content, and thus effect the *206/238* and *207/206* ratios. This factor has been the subject of a recent investigation completed at this laboratory [32]. It is possible to measure the radon leakage from uranium-bearing minerals with good reproducibility. The results show that samarskites leak to an extent of less than 0.1%, uraninites and pitchblendes leak from 0.1% to 10%, and secondary uranium minerals such as carnotite may leak 20% or higher. It has also been found that the leakage rate is temperature dependent, as expected for a diffusion process. The leakage is also dependent to a secondary extent on particle size. Since it can be shown theoretically that radon cannot diffuse through a perfect crystal lattice, the leakage is related to the submicroscopic fissures which

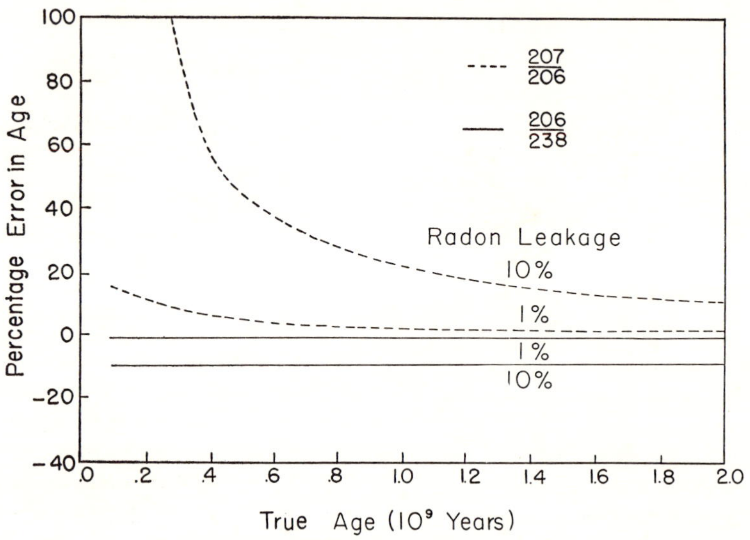

FIG. 1. The effect of common lead on *207/206* and *206/238* ages. A and D no common lead correction. B, C and E, F show case of maximum uncertainty in choice of isotopic composition used in making the correction.

determine the total effective surface area in the mineral. This is dependent on the amount of ionic substitution as well as the age and the tectonic history of the mineral.

Figure 1 shows the percentage error involved in the *207/206* and *206/238* ages for 1% and 10% radon leakage for minerals of various ages. It is observed that the error in the *206/238* age is directly proportional to the percent radon leakage regardless of the age of the mineral, whereas the error in the *207/206* age increases exponentially as the age of the mineral decreases.

Experiments have demonstrated that radon leakage is universally present in uranium minerals and is frequently of sufficient magnitude to affect the measured age. Since it is temperature dependent, the integrated radon leakage (over the history of the mineral) is difficult to estimate, so that an appreciable uncertainty from this source remains in the *206/238* and *207/206* ages. It is noteworthy that the *207/235* age is independent of radon leakage. Also, the *206/210* age is independent of radon leakage if the rate of leakage has been nearly constant.

2.4. Common Lead

One of the serious problems of the lead method is the correction for common lead. For example, Table VII shows the effect of common-lead correction on the crude age of a mineral. The Wilberforce specimen has

TABLE VII. Effect of common lead on "crude" Pb/U Ages.

Locality	Pb/U ("crude")	Lead isotope ratios—Ages (million years) *206/238*	*207/235*	*207/206*
Wilberforce, Ont.	1060 ± 10 Common lead less than 0.5%	1077 ± 20	1050 ± 20	1035 ± 30
Caribou Mine, Colo.	740 ± 10 (True age about 25 million years) Common lead 97%	23 ± 10		

negligible common lead and hence gives a reliable "crude" age, whereas the Caribou Mine specimen would be in error by 700 million years if this correction is ignored.

The isotopic compositions for some common leads reported by Collins *et al.* [33] are shown in Table VIII. The first two were chosen to show the maximum range. This amounts to ±25% of the mean for *206*, ±6% for *207* and ±7% for *208*. It will be seen from the Broken Hill and Casapalca samples that leads from the same mine or restricted locality have essentially the same isotopic composition regardless of the lead mineral examined.

The sets of samples from the Joplin Lead-Zinc District, the Connecticut pegmatites and the North Carolina pegmatites show that differences in a metallogenic district are not large [33a, 33b]. The last group in Table VIII gives the average of common-lead compositions published by Nier [5, 6]. These averages show a constant trend but individual areas may differ greatly from the average, depending on the geochemical history.

The error in the age of a mineral due to the variations in common-lead content depends on the concentration of common lead and the choice of the proper composition of that lead.

TABLE VIII. Isotopic abundances of some related specimens of "common" lead.

No.*	Locality	Isotopic abundances referred to Pb Pb^{204} = 1.000		
		206	*207*	*208*
C-14	Rosetta Mine, Barberton, South Africa	12.65	14.27	32.78
C-12	Worthington Mine, Sudbury, Ontario	26.00	16.94	45.57
C-2	Kengere, Belgian Congo	18.94	15.86	39.70
C-15	Great Bear Lake, N.W.T., Canada	15.93	15.30	35.3
C-16	Ivigtut, Greenland	14.65	14.65	34.48
C-17	Joplin, Missouri	22.35	16.15	41.8
V-23	Joplin, Missouri	22.42	16.01	41.5
V-24	Joplin, Missouri	21.78	15.72	40.8
N-9	Joplin, Missouri	21.65	15.88	40.8
N-10	Joplin, Missouri	21.65	15.74	40.4
N-11	Joplin, Missouri	22.38	16.15	41.6
N-22	Broken Hill, N.S.W.	16.07	15.40	35.5
N-23	Broken Hill, N.S.W.	15.93	15.29	35.3
N-1	Galena, Casapalca Mine, Pa.	18.85	15.66	38.6
N-2	Bournonite, Casapalca Mine, Pa.	18.67	15.45	38.2
L-10	Middletown, Conn.	18.81	15.85	39.0
L-11	Darien, Conn.	18.87	15.66	38.6
L-12	Roxbury, Conn.	18.78	15.88	39.0
L-26	Silver Hill Mine, N.C.	18.41	15.85	38.3
N-18	Yancey Co., N.C.	18.43	15.61	38.2
N	Post Paleozoic Av.	18.94	15.69	38.64
N	Paleozoic Av.	17.84	15.53	37.90
N	Pre-Cambrian Av.	15.70	15.16	35.25

* *N*—Nier [5, 6]
C—Collins [33]
V—Vinogradov [33a]
L—Lamont [33b]

Figure 2 shows the percentage error found in the age for minerals of various ages with a fixed initial ratio of common lead to uranium of 8 mg Pb to 1 gm U. The graph shows the effect on the *207/206* and *206/238* ages for the case of (1) no common-lead correction-curves *A*, *D*, and (2) common-lead correction made, but with maximum uncertainty in the choice of isotopic composition used in making the correction. The first case applies to all of the pre-1938 uranium–lead ages. The second case applies to any district where no common lead has been analyzed. It is clear that the effect increases exponentially as the true age of the mineral becomes younger.

The effect of various percentages of common lead on the *207/206* ages is shown in Fig. 3. In this graph two cases are illustrated, one for the maximum error in the common-lead correction (i.e., complete uncertainty

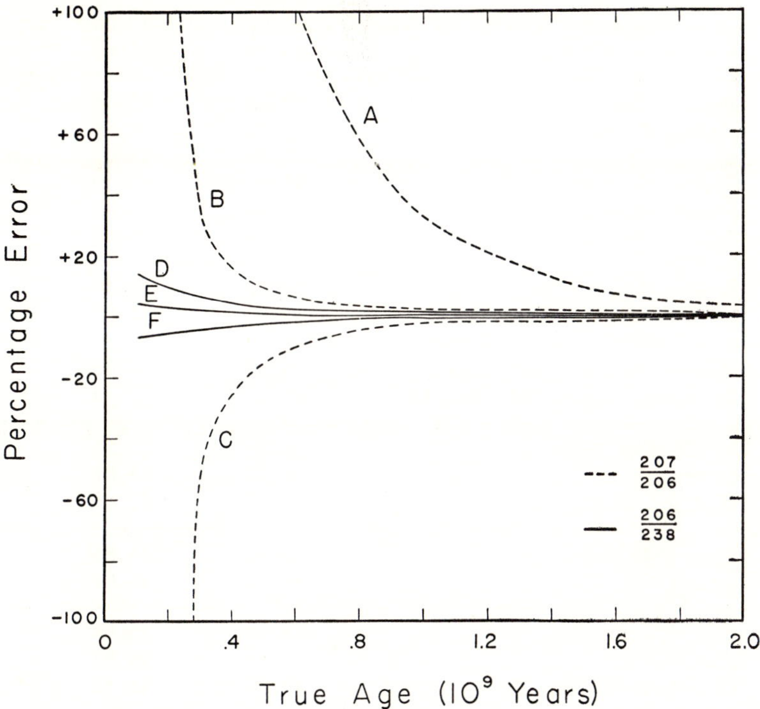

Fig. 2. Percentage error in *207/206* and *206/238* ages for 1% and 10% radon leakage for minerals of various ages.

as to the common-lead isotopic composition to use) and the other for the error expected if the variation within a district has been defined. This emphasizes the importance of measuring many common leads in the region surrounding an occurrence of uranium mineralization.

The common-lead correction is generally based on the nonradiogenic Pb^{204}. Since the relative abundance of this isotope is subject to the largest error (1% or greater), however, it is preferable to use the Pb^{208} as a basis for the common-lead correction, provided that there is negligible thorium present in the mineral.

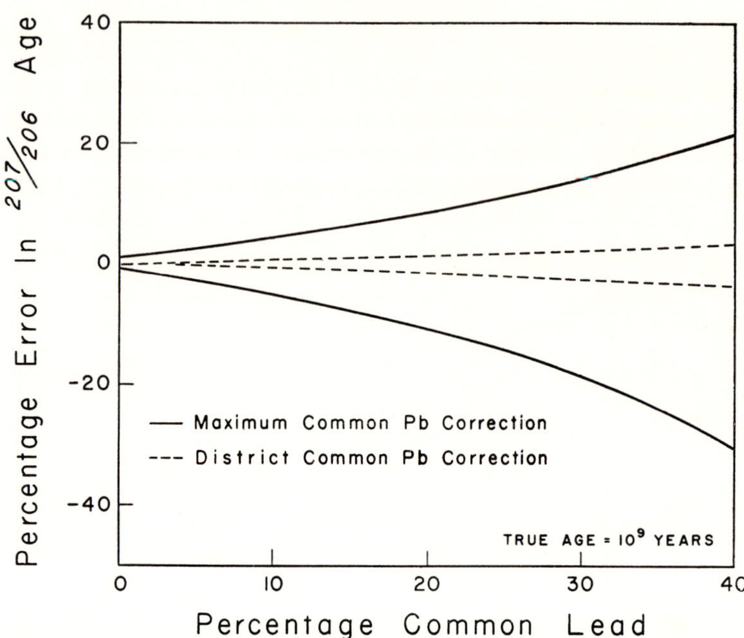

FIG. 3. Effect of various concentrations of common lead on *207/206* ages. Two cases are shown: (1) maximum uncertainty in isotopic composition of common lead used for correction; (2) uncertainty in isotopic composition of common lead limited to range of compositions observed in a district.

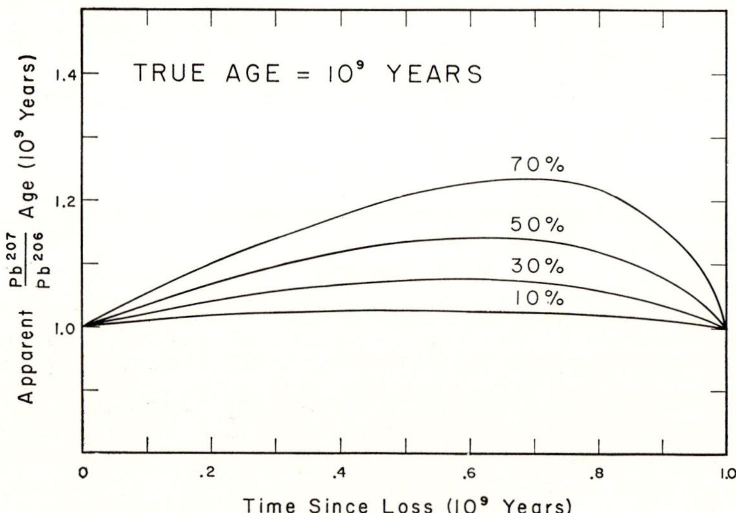

FIG. 4. Error in *207/206* age due to leaching of uranium at a specific time for a mineral 1×10^9 years in age.

2.5. Leaching and Alteration

If any of the elements in the decay chain from uranium or thorium to lead are leached out preferentially during the history of the mineral, the age may be affected. If intermediate members are leached for a relatively short period the effect will be negligible except on the *206/210* ratio.

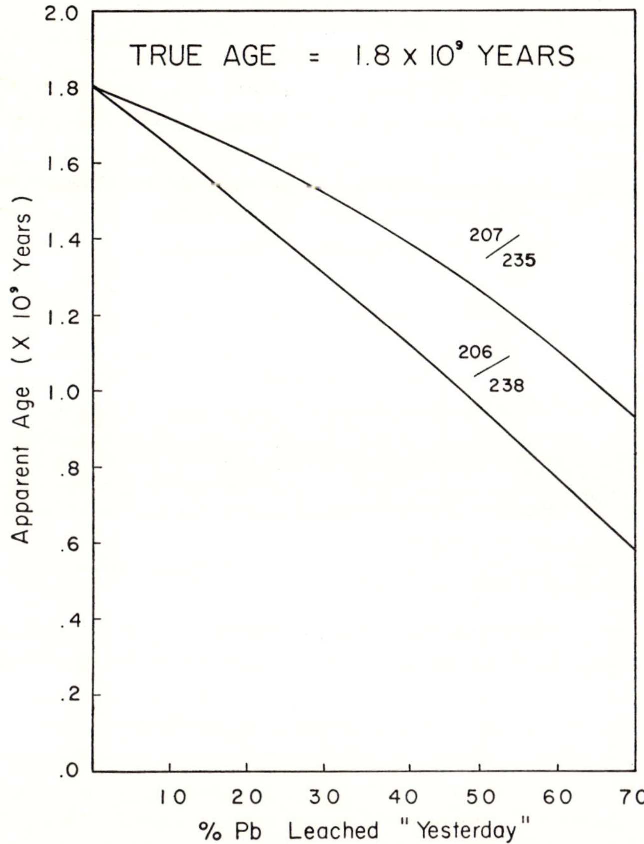

FIG. 5. The effect of lead leaching on the *206/238* and *207/235* ages.

It is required to know the leachability of uranium, thorium, and lead under various ground water conditions. It is unlikely that the alteration of a radioactive mineral occurs while the mineral is deeply buried, unless the area has undergone subsequent regional metamorphism. Therefore the critical situation occurs while the mineral is close to the surface, subject to active ground water circulation and alteration. This probably is most significant during the past 10^4 to 10^6 years. Unfortunately, there is no direct experimental evidence on the leachability of uranium and lead by ground water. Phair and Levine [34] showed that uranium is leached preferentially to lead in pitchblende in dilute sulfuric acid solutions, but this is a different situation than in the case of ground water alteration. Starik [35] reported that ground water did leach radium preferentially to uranium, but did not mention lead. An investigation of this problem has been initiated recently at the Lamont Observatory.

If uranium is leached preferentially to lead during recent weathering, the *206/238* and *207/235* ages will be high and the *207/206* or *206/210* ages will be unaffected (Fig. 4). If radium is leached preferentially to uranium, thorium, and lead during the past 10,000 years, the *206/238*, *207/235*, and *207/206* ages will be unaffected, but the *206/210* age will appear too old. On the other hand, the recent leaching of lead preferentially to uranium will have the opposite effect on the *206/238* and *207/235* ratios (Fig. 5). To obtain a complete understanding of the validity of an age derived from one of these isotopic ratios, it is necessary to determine the extent of inequilibrium in the decay series by direct measurement of uranium, radium, Pb^{210}, and total lead. This has not been done in previous studies of the lead method.

The data of Table VI show that in certain cases the extent of alteration is negligible. However, samples *N*-28, *N*-17, and *N*-29 indicate that preferential uranium leaching has occurred, and samples *N*-14, *C*-30, *C*-79, and *C*-89 suggest the leaching of lead. It is clear that further work is required.

2.6. Errors Due to Other Factors

The error in chemical analysis is generally only 0.1 to 1.0% for U, Th, and Pb. If the lead concentration is low, as in Tertiary minerals, the errors can become quite large, although with special procedures and considerable care the error in the analysis should not exceed a few percent.

The mass spectrometer error of 0.2 to 0.5% becomes significant only if all other variables are essentially eliminated. If the common-lead content is low the ultimate precision of the lead method would approach 0.5%. Except for the need of some improvement in the precision of measuring the scarce *204*, the mass spectrometric technique is not presently a limiting factor in the accuracy of the lead method.

The uncertainty in the half-life of U^{235} is nearly 2% [36]. This contributes only 1% error to the *207/235* ratio, but the *207/206* age is much more sensitive to this factor.

2.7. General Discussion

In most of the samples shown in Table VI, the *206/238* age is generally a little lower than the *207/235* age, but the *207/206* age is considerably higher. This can be explained in various ways: (1) Radon leakage would have the effect of lowering the *206/238* age while raising the *207/206* age. If this is the only source of error, the *207/235* age would be the most reliable age. (2) Leaching of lead late in the history of a mineral would produce the same effect. (3) If the natural half-life of U^{235} is larger than reported, the age for *207/235* would be high. The *207/206* age would be even higher. (4) Excessive neutron capture by Pb^{206} would produce the observed result.

The last possibility proves to be highly unlikely on perusal of the available data on neutron capture cross sections. The fact that virtually all of the *207/206* ages lie to the right of the curve drawn for the "best" half-life suggests that this factor may contribute to the variation in the ages. However, since radon leakage and lead leaching would displace the *207/206* ages in the same direction, the answer is not obvious. Certainly if the half-life were the major factor, the anomalies would be expected to be quite regular. This is not the case.

If the radon leakage of a given mineral is measured quantitatively in the laboratory and this correction applied, the ages derived from the various isotopic ratios should be brought into agreement providing no other effect is present. If the *206/238* ratio is still low and the *207/206* high compared with the *207/235*, it suggests lead leaching and that the revised *207/206* age is the correct one. If the radium content of the mineral is low, this is further confirmation of recent alteration. The *206/210* age may be high or low depending on the total quantity of lead leached out before "yesterday" and the ratio of actual radium to equilibrium radium in the mineral at the time of analysis.

The following conclusions can be drawn concerning the reliability of the various lead ratios. If all ratios agree, the age is correct. If the ratios differ even after correction for radon leakage, and if the mineral is at least one billion years old, the corrected *207/206* age is the best.

For younger minerals the problem is more complicated. The degree of inequilibrium must be estimated from the concentrations of the intermediate members of the series. This in turn may be used to interpret the most probable age. For very young minerals the *206/238* age is the most probable.

The *206/210* ratio is not affected by the leaching of uranium in the relatively recent past, and yet is quite precise for young as well as old minerals. This makes the *206/210* ratio a more reliable age index than the *207/206* ratio for minerals whose ages are less than one billion years, providing leaching has not occurred. The experimental techniques and theory of the Pb^{210} measurement have been described by Kulp, Broecker, and Eckelmann [37].

At the present time there are only a few localities for which the age may be considered as established within 5%. These include Great Bear Lake, 1400; Wilberforce, Ontario, 1070; Parry Sound, 1030; Besner, 780; Katanga, Belgian Congo, 610; Bedford, New York, 350; Middletown, Conn., 260; and Gilpin Co., Colorado, 60 million years. There are many others which are probably correct to 20%, but these are being or will be redetermined shortly along the lines discussed in this paper, so that accuracy comparable to these others will be attained.

3. The Carbon-14 Method

The discovery of natural radiocarbon by Libby and co-workers [38] produced the most significant advance in geochronometry since the discovery of radioactivity. The relatively short half-life (5568 yr) opened the possibility of dating during the last few tens of thousands of years over which there exists such detailed geological and archeological information. The number of laboratories making measurements by this method is increasing exponentially with consequent improvement in techniques. The pioneer work of Libby is summarized in his concise book entitled *Radiocarbon Dating* [39]. The sets of dates published by the summer of 1954, originating from the laboratories at the University of Chicago, Yale, and the Lamont Geological Observatory (Columbia), have appeared in *Science* [40–45]. Interpretations of these data are found in a wide variety of scientific journals.

3.1. Principles

Cosmic ray primaries cause nuclear reactions in the upper atmosphere which produce neutrons. These neutrons are largely used up in the reaction $N^{14}(n,p)C^{14}$. The C^{14} atoms thus produced are rapidly oxidized to carbon dioxide which becomes thoroughly mixed in the atmosphere in much less than a century. This radioactive carbon dioxide enters the photosynthetic cycle of plants, animals and air, and equilibrates with the carbonate in surface ocean water. If it is assumed that the integrated cosmic-ray flux is constant, the concentration of C^{14} in any part of the atmosphere-biosphere-hydrosphere should be constant. Minor variations due to fractionation effects may occur but can be measured by C^{12}/C^{13} studies. This concentration has been found to be about 15.3 dpm per gram of carbon in living plants and animals. It is similar for shell and air.

If a substance is removed from the photosynthetic cycle, as by death for an animal, it no longer takes in any C^{14}, so that the equilibrium concentration during life decays away at the known half-life (5568 ± 30 yr). Therefore the ratio of the specific activity of any ancient carbon-bearing sample to the modern or "living" value will be a measure of the time since death or isolation.

This theory has been checked experimentally by calculating the age by the natural radiocarbon measurement and comparing it with the known age from history, tree rings, or other methods. Figure 6 shows a plot of

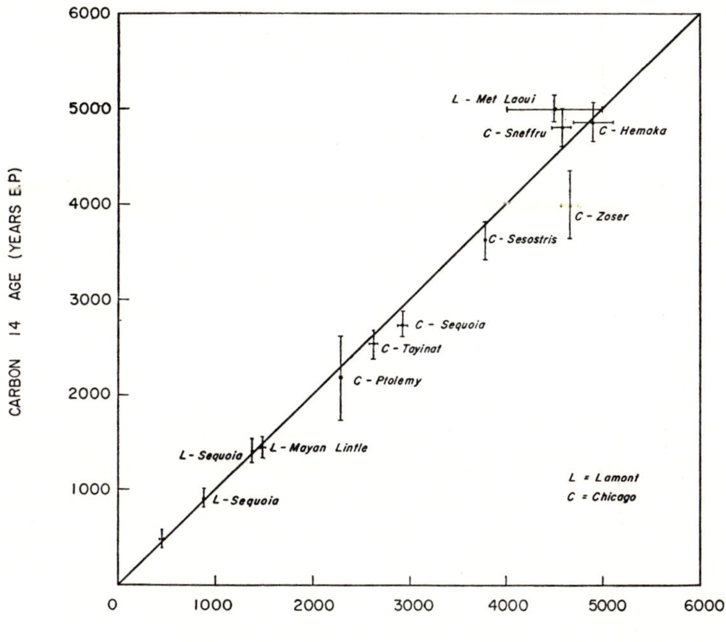

FIG. 6. Plot of carbon-14 vs. historical ages for selected samples.

some of these calibration samples. For some of the archeological samples beyond written history the estimated uncertainty is much larger than for the more recent samples.

3.2. Assumptions

There are two important assumptions in the carbon-14 method which are subject to experimental verification. One is that the C^{14} produced in the upper atmosphere is mixed in the entire photosynthetic cycle rather rapidly compared to 100 years (the present practical limit of measurement). The other is that the integrated cosmic ray flux has been constant.

That mixing has been complete, at least with the present experimental error, is well established. Libby [39] gives a table of the C^{14} concentration in living trees of many types from widely different latitudes and altitudes. All of the results fall within about two standard deviations (95% confidence level). Kulp and Carr [46] present data which further show the constancy of the C^{14} concentration in trees, shells, ocean water, and carbon dioxide in the air. It now appears that mixing is complete to 5%.

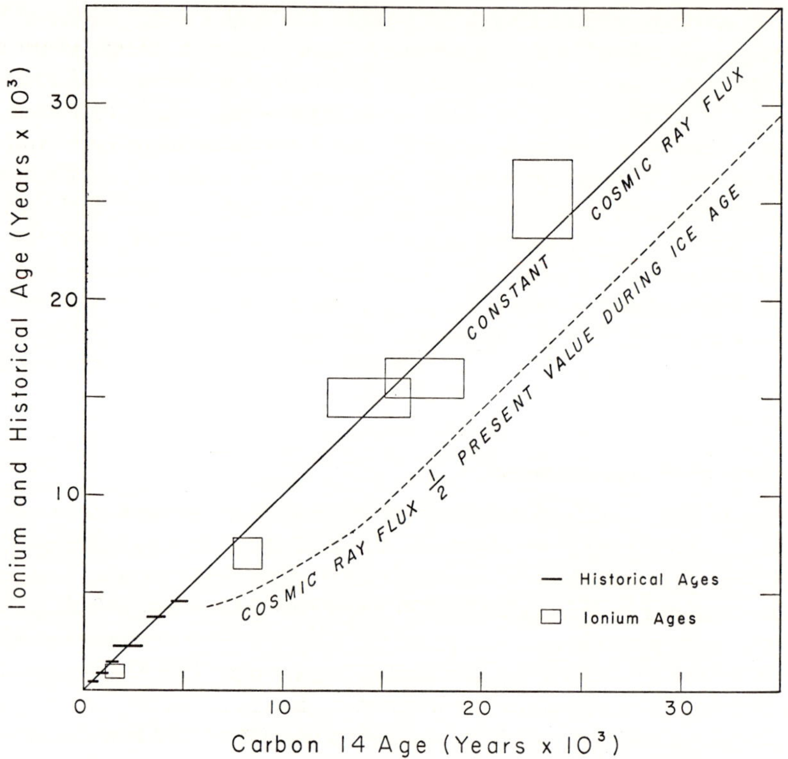

FIG. 7. Correlation of carbon-14 age of various samples with age determined by written history or the ionium method.

Greater precision in the measurements may ultimately reveal local anomalies.

Figure 7 shows the ages obtained on certain layers of deep sea sediments by both the carbon-14 and ionium methods of age determination. Since the ionium method is independent of the cosmic ray flux but the carbon-14 method is directly dependent on it, these data show that the integrated cosmic ray flux has been constant to within 5–10% over the past 30,000 years. If the cosmic ray flux was one half its present value during the ice age, the points should follow the dashed curve. Further work is being done in the laboratory to make this study more precise.

3.3. Experimental Developments

The original technique of measuring natural radiocarbon as developed by Libby [39] consisted of counting solid carbon mounted inside a Geiger tube. Elaborate shielding was necessary, which consisted of eight inches of iron to reduce natural gamma ray and soft cosmic ray components, and an anticoincidence ring to eliminate the high-energy mesons. This technique permitted age measurement back to about 20,000 years ago. The first major improvement was the addition of a mercury shield inside the anticoincidence ring by Kulp and Tryon [47], which extended the range to 30,000 years. Subsequently many methods have been and are being studied to measure natural radiocarbon. A summary of the status in early 1954 is given in Table IX. In general it can be concluded that the solid carbon-Geiger counting method is now superseded, that the ion chamber method never had a chance, that routine measurements in the future will probably be done by proportional counting of gases, but that the maximum sensitivity, range, and accuracy will be obtained by liquid scintillation counting. This is a very active field of research at the present time.

3.4. Geological Applications

One of the most important geological problems is the time of retreat of the last continental ice sheet. In Table X the first two samples of extinct forest now overlain by the latest drift show that the last continental glacier was still in Wisconsin 11,000 years ago. Many additional samples are making it possible to understand the absolute time of the advance and retreat of the ice during the last part of the ice age. The Mississippi delta and Bermuda samples show the worldwide change in sea level accompanying the last glacial period. The weathered surface at 273 feet corresponds to the maximum sea level lowering. Then, as the ice melted, there is a complete record of sea level rise as a function of time. It was about 80 feet lower than now 11,000 years ago, 70 feet 9,000 years ago, 50 feet 7,000 years ago, and 25 feet 3,000 years ago.

Several samples of peat at the bottom of the peat section from different parts of the Florida Everglades have been dated. They all give ages around 5,000 years, suggesting that just prior to this, the lower peninsula of Florida was under water. It is important to note that this apparently occurred near the thermal maximum.

On the California coast, marine terraces are very well defined. It was of interest to find out if the lowest, last terrace was produced within 30,000 years. It apparently was not.

The last great eruption of La Soufriere, Island of Guadeloupe, was

TABLE IX. Summary of techniques being used to measure natural radiocarbon.

Counter type	Lab.	Form	Bkgd. elim.	Eff. (%)	Sample weight (gm)	Modern count (cpm)	Bkgd. count (cpm)	Error on 5600 year old sample (2-day count)	Age Limit (2-day count)
Geiger screen wall	Chicago	C	Steel, A-C*	6	6.0	5.4	5.0	450	28,000
	Columbia	C	Steel, A-C	6	6.0	5.4	3.0	450	28,000
	Denmark	C	Steel, A-C	6	6.0	5.4	4.0	400	30,000
Liquid scintillation (photomultiplier tube)	Chicago	Alcohol	Coincidence, discriminator, mercury, deep freeze	40	15	90	20	60	50,000
	Manitoba	Toluene	Lead, discriminator	80		150	100	50	52,000
	Columbia	Benzene	Coincidence, discriminator, mercury, deep freeze	40	80	480	20	15	65,000
	(Future)	Benzene	Coincidence, discriminator, mercury, deep freeze	80	400	4,800	50	5	80,000
Cloud chamber	Brookhaven	CO_2	Visual, underground	25	66	200	10?	—	—
Ion chamber	Calif. Inst. Tech.	CO_2	Underground	—	1.5	4×10^{-17} Amps	5×10^{-17} Amps	800	15,000
Geiger	Michigan	CO_2	Steel, A-C, transformer oil	50	0.26	2.1	3.0	700	24,000
Proportional counter	Netherlands	CO_2	Steel, lead, A-C	70	0.48	5.6	3.0	300	32,000
				70	2.4	28.0	3.0	100	43,000
	Argonne	CO_2	Steel, A-C	60	0.78	5.7	25.0	700	20,000
	Magnolia	CH_4	Steel, A-C	80	0.86	10.4	14.4	300	28,000
	USGS	C_2H_2	Steel, mercury, A-C	70	1.07	10.4	2.5	150	36,000
	Columbia	C_2H_2	Steel, mercury, A-C	70	2.9	28.1	8.1	100	41,000
	(Future)	C_2H_2	Steel, mercury, A-C	70	40.0	235	22.5	45	58,000

*Anticoincidence

TABLE X. Geological samples.

Sample No.	Description	Age (years before present)
C-308	Two Creeks, spruce wood	10,880 ± 740
C-366	Two Creeks, peat	11,100 ± 600
L-125*I*	Mississippi delta, weathered surface, 273 ft	Older than 30,000
L-125*G*	Mississippi delta, shell 73 ft	9,000 ± 200
L-125*A*	Mississippi delta, wood 25 ft	2,900 ± 300
L-111*A*	Bermuda drowned forest, 70′–90 ft	11,500 ± 700
L-111*B*	Bermuda peat	6,900 ± 150
L-141*A*	Florida Everglades peat	4,900 ± 200
L-114*A*	Southern California, 75 ft (lowest) terrace	Older than 30,000
L-102*A*	La Soufrière, charred wood from last eruption	550 ± 150
L-101*B*	Eagle River, Alaska, peat	14,300 ± 600
L-150*C*	Rapa Island, Central South Pacific lignite	Older than 30,000

dated at only 550 ± 150 years ago by a piece of charred wood underneath the volcanic deposits. The Eagle River peat shows the existence of a local warm phase in Alaska about 14,000 years ago, prior to the end of the Wisconsin glaciation. Apparently several tens of thousands of years are required to form lignite from peat, as noted for the youngest lignite on Rapa Island.

A perennial problem of geochemistry and geology has been the question of the origin of petroleum. Most theories included a long period of time—e.g., 1,000,000 years—for the hydrocarbon molecule to be prepared. This was followed by migration and accumulation. It had further been frequently stated that Recent marine sediments contained no hydrocarbons. P. V. Smith [48] of the Esso Laboratories, began a new search for such hydrocarbons and discovered them in appreciable quantities. Not only did he find enough hydrocarbons to account for known petroleum reserves, but the spectrum of compounds was similar to crude oil. The problem remained as to whether these hydrocarbons had diffused up from oil-bearing horizons below, or whether they really form *now*. Although the extremely small quantities of hydrocarbons available required the development of special techniques, it was possible by Carbon 14-determinations to demonstrate that the hydrocarbons were geologically Recent, and thus the problem of the origin of oil becomes the problem of accumulation only.

Carbonate from ocean water at various depths is being assayed at the Lamont Observatory for C^{14} as a possible means of defining rates of ocean circulation. Some preliminary results are shown in Table XI. Although it is too early to make any detailed analysis, it does appear that there is

TABLE XI. Ocean water measurements.

Location	Depth (ft)	Apparent age (yr)
13°35′N, 66°35′W	Surface	Recent
19°24′N, 78°33′W	Surface	Recent
41°00′N, 54°35′W	Surface	Recent
38°42′N, 67°54′W	13,500	450 ± 150
63°46′N, 00°26′W	10,440	500 ± 200
34°56′N, 68°14′W	16,560	1,550 ± 300
35°46′N, 69°05′W	15,300	1,950 ± 200
58°19′N, 32°57′W	6,000	1,600 ± 130
53°53′N, 21°06′W	9,100	1,900 ± 150

a significant difference between surface and deep water. The first deep sample was taken at the edge of the continental shelf and the second in the far north. Both of these might be expected to be younger than the deep mid-Atlantic ocean water.

4. The Strontium Method

The decay of Rb^{87} to Sr^{87} as a possible method of age determination was first suggested by Goldschmidt [49] in 1937, but it was Ahrens [50] who did the pioneer experimental work in an attempt to establish the method.

4.1. Principles

Rubidium consists of two isotopes, Rb^{87} (27.2%) and Rb^{85} (72.8%). The Rb^{87} is beta active with an apparent half-life of 6.3×10^{10} years, and thus decays to Sr^{87} which is present in ordinary strontium to the extent of about 7%. Therefore a mineral that is relatively rich in rubidium (0.1 to 2%) and low in ordinary strontium (less than 0.1 to 0.01%) can be dated. The long half-life indicates that the method will be primarily useful for older minerals. The primary application of the method has been to lepidolite (lithium mica) and amazonite (potassium feldspar), but it is possible to utilize biotite and thus open up all granite and most metamorphic rock to dating.

The accuracy of a Rb-Sr age determination depends on a number of factors, including the physical constants, the isotopic composition of rubidium and strontium, the analytical techniques employed to determine the ratio Sr^{87}/Rb^{87}, and the environment of the mineral.

4.2. *Results*

The strontium method has had a curious history. The early work of Ahrens [50] was done with the emission spectrograph, a technique which is particularly subject to systematic error of absolute concentration. The ages (in million years) obtained by the strontium method using this technique on a number of classic localities, Pala (~120), Maine (~280), New Mexico (~800), Brown Derby (~850), "old granite," S.E. Africa (~900), S.E. Manitoba (~2250), seemed to agree rather well with the available lead ages, although in no case had a complete lead isotopic study been made in these areas. On the other hand, some of Ahrens' dates, such as Middletown, Conn. (~2500), were clearly too high, and the Black Hills date (~900) was too low. It was concluded, however, that the strontium method ages agreed with the lead ages, and since the latter were considered accurate, the strontium method appeared established.

In the last few years a number of laboratories, notably the Department of Terrestrial Magnetism of the Carnegie Institution, have started investigations of strontium using isotope dilution techniques [51].

Taking strontium as an example, a known quantity of the rare isotope Sr^{84} is added to the sample as a "spike." The strontium separated roughly from the sample is put on a filament of a mass spectrometer, and the ratios of the strontium isotopes are used to determine the original concentration of strontium in the sample. This technique appears to yield reproducible absolute values for the Sr^{87}/Rb^{87} ratio to about 2% compared to the 10–20% error inherent in the emission spectrograph technique.

The ages obtained by the isotope dilution technique, however, appear abnormally high. Table XII gives a comparison of the ages obtained from various localities by these different methods. According to theory, the analysis performed by the isotope dilution technique should be superior

TABLE XII. Comparison of strontium and lead ages from selected localities.

Locality	Lead age or best geologic age (million years)	Sr^{87}/Rb^{87} age (million years)	
		Emission spectrograph (Ahrens, 52)	Isotope dilution (Davis and Aldrich, 51)
Pala, Calif.	130	120	150
Middletown, Conn.	260	500	—
Brown Derby Mine, Colo.	900	850	2000
Black Hills, S.D.	1450	900	2000
S.E. Manitoba	2400	2250	3360
Pope's Claim, S. Rhodesia	—	2150	3740

in every way to that of the emission spectrograph. If this is the case, there is a major error somewhere. The interlocking of the lead isotope ratios makes the lead age less suspect than the strontium age. The most likely sources of error which would make the strontium ages high are the assumed half-life or leaching of rubidium from a mineral. Unfortunately, at present there are not enough localities which have been analyzed completely for uranium–lead and rubidium–strontium to determine whether or not the strontium ages are higher by a constant fraction. If the anomalies are not systematic the leaching of rubidium would appear to be the best explanation. This conclusion should be followed by experiments to determine the conditions for leaching.

It is clear that although the strontium method is a promising one, there are serious problems yet to be solved before it can be applied to quantitative time measurement.

5. The Potassium Methods

5.1. *Principles*

K^{40}, which occurs in natural potassium to the extent of 0.0119%, decays by beta decay to Ca^{40} and by *K*-capture and de-excitation to A^{40}

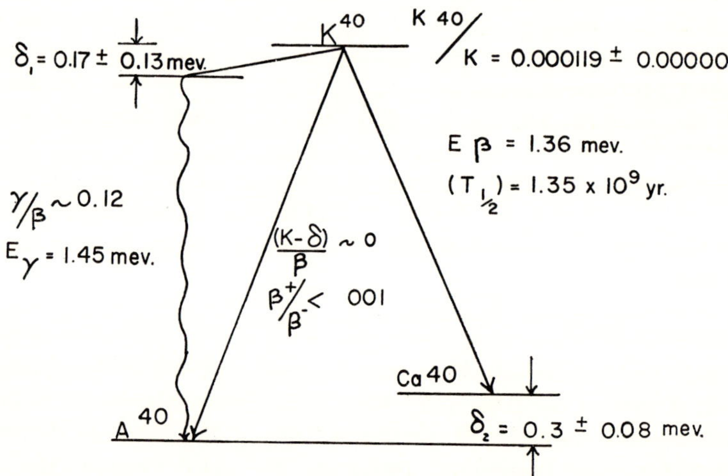

Fig. 8. Decay scheme of K^{40}.

according to the decay scheme shown in Fig. 8. Most of the decay is to Ca^{40} with a half life of 1.35×10^9 yr.

This scheme is potentially the most important method of determining geological time since potassium is a common element in most important rock types, in contrast to the relatively rare elements uranium, thorium, and rubidium. Moreover, there are possibly two methods. Ahrens and Evans [52] have outlined the possibilities for the Ca^{40} decay. It is not as promising as the A^{40} decay since the concentration of K^{40} is low. Ca^{40} is the common isotope of ordinary calcium and there are appreciable (0.01–0.1%) quantities of ordinary calcium in virtually all potassium minerals. However, with precise solid source mass spectrometry, it may be possible to utilize the K^{40}–Ca^{40} decay.

Most of the effort in recent years [53–57] has been placed on the K^{40}–A^{40} decay since there is essentially zero A^{40} in an igneous mineral at the time of its formation, and A^{38} is available as a "spike" for isotope dilution analysis.

There are a variety of problems to be solved before this clock becomes useful. Until a few years ago even the beta half-life was uncertain. Once this was defined, the branching ratio posed, and is still posing, the largest problem. In addition to this matter of the physical constants, there is also the difficulty of releasing all of the argon from a mineral at the time of fusion and the possible leakage of argon from the lattice during the lifetime of the mineral.

From studies on radon leakage carried out at this laboratory [32], it would be predicted that the only argon likely to leak would be that produced by the decay of K^{40} atoms at the surface of the mineral. Since the surface area of most common potassium minerals is very low, it does not appear probable that argon loss is important, unless the temperature of the mineral approaches 400–500° C where diffusion becomes appreciable.

It also appears that the problem of releasing all of the argon has now been solved. If either direct fusion or NaOH flux is employed the argon appears to come off quantitatively. With a sodium metal flux this was not the case.

5.2. Branching Ratio

During the past decade the branching ratio A^{40}/Ca^{40} has been estimated all the way from 0.05 to 1.9 by various investigators. Because this quantity must be known with precision for any absolute age determination, this is a very immediate problem. We shall consider the direct physical methods of obtaining this quantity first.

The gamma radiation associated with K^{40} decay has been assigned to the K^{40}–A^{40} branch for several reasons: (1) beta and gamma measurements and mass measurements have established that the energy of the gamma radiation is greater than the difference in energy between the K^{40} and Ca^{40} ground states (2) no β–γ coincidences have been observed, (3) the coincidences which have been observed can be associated with the γ radiations and the products of K-capture. Positrons have been looked for and not found. Thus, if only direct physical measurements by counting techniques are considered, the activities are in reasonably good agreement and indicate a branching ratio of 0.12 ± .02.

A chemical determination by Inghram *et al.* [54] which involved the direct measurement of the ratio of A^{40}/Ca^{40} in an ancient sylvite gave a value of 0.13 ± .01.

The University of Toronto group assumed the ages of four feldspar minerals [55] from lead *207/206* ages and computed the branching ratio from their A^{40} and K^{40} contents. This gave a branching ratio of 0.06. Later [56] they revised this upward to 0.09 when they discovered incomplete release of argon in the earlier work. Wasserburg and Hayden [57] measured the A^{40} in a feldspar that had a *207/235* age of about 0.76×10^9 years. The results are consistent with a branching ratio of 0.12 ± .02. Later unpublished work by the same authors appears to give a lower ratio.

It may be concluded that the potassium method still needs a great deal of study. The fact that a number of competent investigators are persistently working on the problem gives hope for success. If the branching ratio can be established to within 5% by direct physical measurements and several ratios determined by the geologic method are shown to be accordant, earth history will be subjected to a most vigorous examination.

6. The Helium Method

The accumulation of helium in a radioactive mineral due to α-decay as an index of geologic time has been the subject of a large number of investigations since the turn of the century. The interest in this lies in its great potential usefulness in geology, as it is applicable, in principle, to all igneous rocks and possibly to many metamorphic rocks. The history of the helium method has certain features in common with the strontium method.

In the mid 1930's it appeared that the helium method gave ages in agreement with lead ages, and thus that its reliability was established [7–9]. Further work revealed that this was not the case, and that the helium ages were generally low. This was ascribed to helium leakage, and efforts were made to establish "helium retentivities" for the common rock forming minerals by Keevil [58] and Hurley [18]. It appeared at this stage of the development that magnetite had a high retentivity (near 100%) and might be the best mineral to use in helium age work [19].

In 1941, after using several hundred rock samples, Keevil [59] concluded that no type of rock or mineral gives consistent results (including magnetite) because of the variable loss of helium. In the same year Hurley and Goodman [18] published an important paper showing the self-consistency of the helium ages of samples of the Palisades sill taken over a wide area, but also apparently showing the variable loss of helium in different mineral constituents of the rock. They concluded that magnetite retained all of its helium, although later work showed some anomalies even with magnetite. Later Hurley [60] reexamined the situation and did some ingenious experiments to show that, in granitic rocks at least, a large fraction of the uranium and thorium exists outside of the mineral grains along grain boundaries. Helium generated from such a source will leak out quantitatively. Therefore, a large part of the problem all along was the incorrectly high uranium and thorium content assumed for the interior of the mineral grains. Preliminary experiments showed that some of the intergranular uranium and thorium could be leached out in the laboratory without appreciably affecting the helium content. Further work may discover a leaching procedure which will remove all of the intergranular radioactivity without affecting the interior of the crystals. The significance of helium loss from the inside of crystals remains to be determined.

Hence the helium, as the strontium and potassium methods, is still in the stage of development despite the many measurements which have been made. Two of the most promising directions to follow for the helium method are: (1) separated minerals of high activity such as zircon, where the surface (intergranular) contamination is small compared with internal activity (in the case of moderately high radioactive content, an estimate of the relative helium leakage may be ascertained from the measured radon leakage), (2) rocks which are low in radioactivity (basalt, diabase) and may not have "hot spots" or appreciable surface contamination.

The helium method remains as one of the potentially most valuable geological chronometers for igneous rocks.

7. The Ionium Method

The age methods previously discussed cover geologic time with the exception of the important time between 1,000,000 and 50,000 years ago. The importance of the ionium method is that it spans a considerable fraction of this interval. It is limited to deep sea sediments in its application and to 400,000 years ago in its range, but since the ocean occupies nearly three-fourths of the surface of the globe these limitations are not too severe.

The ionium method was first suggested by Petterson [61] but was first studied experimentally by Piggot and Urry [21]. More recent studies have been made on the mechanics by Holland and Kulp [62, 63], and on the assumptions and limitations of the method by Volchok and Kulp [64].

Ionium (Th^{230}) occurs in the decay series of U^{238} (Table II) and has a half-life of about 80,000 years. Thorium is chemisorbed preferentially to radium and uranium on the surface of deep sea sediment particles as they fall through the water column to the bottom. Further, the radium is chemisorbed preferentially to the uranium. Therefore when the particle reaches the ocean floor it has a large excess of ionium, and a small excess of radium on its surface compared to uranium. If homogeneous sedimentation occurs the ionium concentration will decay regularly with

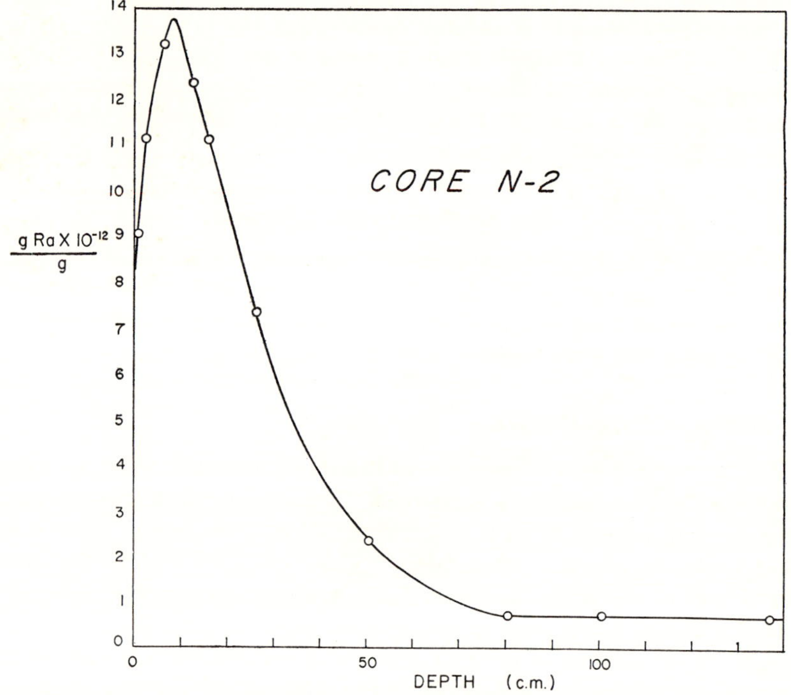

FIG. 9. Concentration of radium as a function of depth in a core.

depth until equilibrium with uranium is reached. Since the half-life of ionium is known, it is then possible to estimate the time since sedimentation for any given layer.

In practice, because of its relative simplicity compared to ionium, the radium content along the deep-sea sediment core is measured. In this case the radium content increases with depth until it equilibrates with the chemisorbed ionium (at about 10,000 years) and then decreases with the ionium. Figure 9 shows an experimental curve from core N-2 taken from the southern Pacific Ocean, which shows these features. For precision dating of the last 20,000 years, it is necessary to make many radium measurements in the upper few centimeters of the core to determine the curvature accurately. Most of the older cores are too thin to yield enough material to satisfy this requirement.

Ocean sedimentation is a complex phenomenon [65] involving redeposition, mixing, and changing sources. The ionium method is only applicable to homogeneous sediment which has been deposited in a "normal" way. This means that the sediment was "rained down," rather than carried in by a turbidity current. In the North Atlantic the average core is a result of such a redeposition process and is quite inhomogeneous. Therefore it cannot be dated by the ionium method.

At the present time the ionium method has been established as a geological chronometer. However, it is limited to the relatively rare homogeneous cores, and care must be taken in the spacing of the measurements to attain suitable precision.

8. Other Potential Methods

In addition to the methods already described, there are several others which do not warrant such extensive treatment, either because they do not appear promising as useful chronometers, or because relatively little work has been done on them.

8.1. Metamict Mineral Method

Radiation damage in certain radioactive minerals such as zircons, samarskite, and euxenite is evidenced by an increasing disorganization of the crystal lattice. Such a mineral is called "metamict." For a given lattice type the amount of damage will be proportional to the specific activity and to the time since mineral formation. As was first pointed out by Holland and Kulp [66], if the specific activity and degree of lattice damage can be measured quantitatively, an estimate of age can be made once a given lattice is calibrated. Using differential thermal analysis to determine the lattice disorder, and α-counting to determine the radiation level, Kulp *et al.* [67] showed that the area under the thermal curve (proportional to the energy released on reordering of the lattice) is linear with alpha activity for samples of samarskite of variable uranium concentration from Spruce Pine pegmatite district, N.C. (Fig. 10). These authors were able to show a rough correlation between known geologic age and the ratio of disorder to specific activity. The difficulty in making this method quantitative is the large effect on the stability of a lattice due to ionic substitution and original crystal imperfection.

More recently Hurley and Fairbairn [68] studied the relationship between crystal damage as measured by the change of the x-ray diffraction angle from the *(112)*-planes and the total alpha-particle irradiation

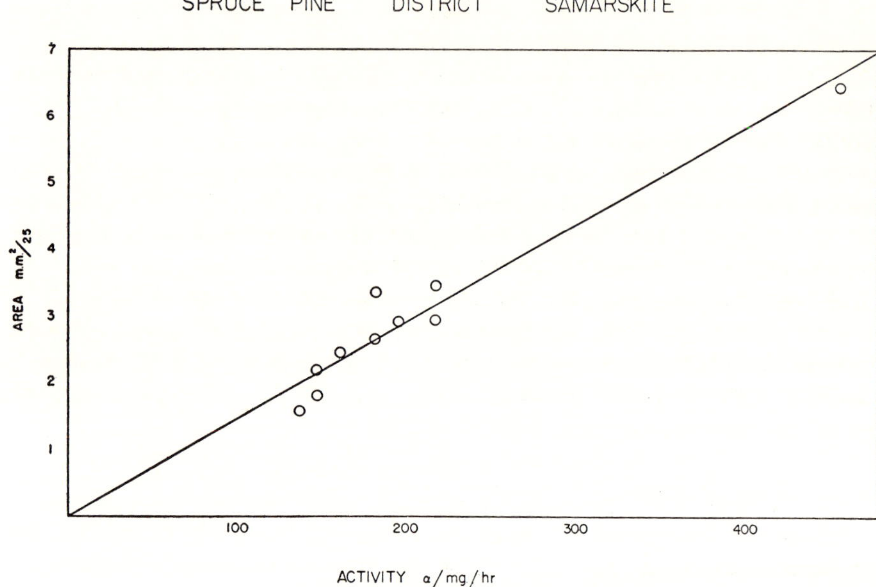

FIG. 10. Area under differential thermal curves vs. alpha activity for Spruce Pine samarskites.

based on age times alpha activity. They are optimistic about the use of this technique for quantitative age measurement for zircons having less than 500 alphas/mg/hr activity.

8.2. Thermoluminescence

Daniels and his co-workers at the University of Wisconsin [69] have examined the effect of very light crystal damage by radioactive bombardment. If a limestone is heated slowly it becomes thermoluminescent over a certain range and with a certain intensity, depending among other things on the radioactive content and the time since the last rearrangement of the crystal structure (age). Therefore, if other variables can be held constant, the ratio of the amount of energy liberated by thermoluminescence to the specific activity should be related to age. Some very rough correlations have appeared from such measurements but, until the complicated effects of ionic substitution, lattice perfection, and temperature are well defined, it is not possible to consider this as a quantitative method for determining geologic age. This method is less promising than the metamict method because of its greater sensitivity to low energy environmental effects.

8.3. Common-Lead Method

The addition of the lead isotopes Pb^{206}, Pb^{207} and Pb^{208} to the earth's crust throughout geologic time due to the decay of uranium and thorium provides another possible age method. By achieving the difficult experimental feat of measuring the isotopic composition of lead in a stony meteorite (primal lead) and the lead in Pacific Ocean red clay (average crust), Patterson *et al.* [70] were able to estimate the age of the planet at 4.7×10^9 years. This is very important in that it represents the first experimental determination of the age of the planet. Farquhar *et al.* [71] have recently proposed that for old galenas (i.e. 2×10^9 or older) it is possible to use the isotopic composition as a quantitative age method. In later periods the continental structure has become so complicated and the lead so remixed that the method breaks down. More study is required before the method can be generally applied.

8.4. Tritium Method

Kaufman and Libby [72] have shown the existence of natural tritium produced by cosmic ray reactions in the upper atmosphere. If the integrated average H^3 content of rainfall over a given area can be ascertained, the decay can be followed for about a century and be used to study the rate of formation of glaciers, movement of ground water, and related hydrologic-geologic problems.

The present phase of the investigation is concerned with a reconnaissance assay of the tritium concentration of surface water. Atomic tests may permit global tracer experiments on some of these problems.

8.5. Spontaneous Fission, Samarium, and Re^{187}

The spontaneous fission of uranium is due almost exclusively to U^{238}. This process may be distinguished from the $n + U^{235}$ fission, in that the U^{235} process produces excessive Xe^{133} and Xe^{134}, whereas the U^{238} process gives Xe^{132} and Xe^{133}. Fortunately, the isotopes of fission xenon have very critical positions near the tops of the yield curves. The half-life for spontaneous fission of U^{238} is quite long (8×10^{15}) but measurable. Macnamara and Thode (1950) have shown how this can be used for age determination. The method would be useful only for uranium minerals.

It is known that samarium has a long-lived isotope with a half-life of about 10^{12} years, but as yet there is no agreement as to its identity. If this is defined, it represents a potential clock in the rare earth minerals. However, the uranium and thorium are generally found where rare earth minerals are present; hence this more elusive clock may not be too valuable.

Suttle and Libby [73] have recently identified Re^{187} as a beta emitter with a half-life of about 10^{11} years. This is also a potential chronometer, but the scarcity of rhenium and the long half-life do not make it appear promising.

Other naturally occurring isotopes will probably be discovered which will be of aid in geochronometry, but it is unlikely that any very abundant ones have been overlooked. There appears to be plenty of work to do on those already in hand.

References

1. Rutherford, E. (1904). Address given to the International Congress of Arts and Sciences. St. Louis.
2. Boltwood, B. B. (1905). The Origin of Radium. *Phil. Mag.* [6] **9,** 599–613.
3. Boltwood, B. B. (1907). The Ultimate Disintegration Products of Uranium. *Am. J. Sci.* [4] **23,** 77–88.
4. National Research Council (1931). Physics of the Earth IV, The Age of the Earth, Bull. No. 80, Washington, D.C.
5. Nier, A. O. (1939). The Isotopic Constitution of Radiogenic Leads and the Measurement of Geological Time. II. *Phys. Rev.* **55,** 153–163.
6. Nier, A. O., Thompson, R. W., and Murphy, B. F. (1941). The Isotopic Constitution of Lead and the Measurement of Geological Time. III. *Phys. Rev.* **60,** 112–116.
7. Urry, W. D. (1933). Helium and the Problem of Geological Time. *Chem. Revs.* **13,** 305–343.
8. Lane, A. C., and Urry, W. D. (1935). Ages by the Helium Method: I Post-Kenweenawan. *Bull. Geol. Soc. Am.* **46,** 1101–1120.
9. Urry, W. D. (1936). Ages by the Helium Method: II, Post-Kenweenawan. *Bull. Geol. Soc. Am.* **47,** 1217–1233.
10. Urry, W. D. (1936). Determinations of the Thorium Content of Rocks. *J. Chem. Phys.* **4,** 34–48.
11. Keevil, N. B. (1938). The Distribution of Helium and Radioactivity in Rocks. *Am. J. Sci.* **36,** 406–416.
12. Keevil, N. B. (1938). The Application of the Helium Method to Granites. *Trans. Roy. Soc. Can. IV* [3] **32,** 123–150.
13. Keevil, N. B. (1942). The Distribution of Helium and Radioactivity in Rocks II. Mineral Separates from the Cape Ann Granite. *Am. J. Sci.* **240,** 13–21.
14. Larsen, E. S., and Keevil, N. B. (1942). The Distribution of Helium and Radioactivity in Rocks III. Radioactivity and Petrology of some California Intrusions. *Am. J. Sci.* **240,** 204–215.
15. Keevil, N. B., Jolliffe, A. W., and Larsen, E. S. (1942). The Distribution of Helium and Radioactivity in Rocks IV. Helium Age Investigation of Diabase and Granodiorites from Yellowknife, N.W.T., Canada. *Am. J. Sci.* **240,** 831–846.
16. Keevil, N. B. (1943). The Distribution of Helium and Radioactivity in Rocks, V. Rocks and Associated Minerals from Quebec, Ontario, Manitoba, New Jersey, New England, New Brunswick, Newfoundland, Tanganyika, Finland, and Russia.
17. Keevil, N. B., Larsen, E. S., and Waul, F. J. (1944). The Distribution of Helium and Radioactivity in Rocks VI. The Ayer Granite-Migmatite at Chelmsford, Mass. *Am. J. Sci.* **242,** 345–353.
18. Hurley, P. M., and Goodman, C. (1941). Helium Retention in Common Rock Minerals. *Bull. Geol. Soc. Am.* **52,** 545–560.

19. Hurley, P. M., and Goodman, C. (1943). Helium Age Measurement I. Preliminary Magnetite Index., *Bull. Geol. Soc. Am.* **54,** 305–324.

20. Evans, R. D., Goodman, R. D., Keevil, N. B., Lane, A. C., and Urry, W. D. (1930). Intercalibration and Comparison in Two Laboratories of Measurements Incident to the Determination of the Geological Ages of Rocks. *Phys. Rev.* **55,** No. 10, 931-946.

21. Piggot, C. S., and Urry, W. D. (1942). Time Relations in Ocean Sediments. *Bull. Geol. Soc. Am.* **53,** 1187–1210.

22. National Research Council (1937 to 1953). Report of the Committee on the Measurement of Geologic Time, 2101 Constitution Ave., Washington, D.C.

23. Larsen, E. S., Jr., Keevil, N. B., and Harrison, H. C. (1952). Method for Determining the Age of Igneous Rocks Using the Accessory Minerals. *Bull. Geol. Soc. Am.* **63,** 1045–1052.

24. Keevil, N. B. (1939). The Calculation of Geological Age, *Am. J. Sci.* **237,** 195–214.

25. Kulp, J. L., Bate, G. L., and Broecker, W. S. (1954). Present Status of the Lead Method of Age Determination. *Am. J. Sci.* **252,** 345–365.

26. Kerr, P. F., and Kulp, J. L. (1952). Pre-Cambrian Uraninite, Sunshine Mine, Idaho. *Science* **115,** 86–88.

27. National Research Council (1949). Report of the Committee on the Measurement of Geologic Time for 1947–48, pp. 18–19.

28. Stieff, L. R., and Stern, T. W. (1952). Identification and Lead-Uranium Ages of Massive Uraninites from the Shinarump Conglomerate, Utah. *Science* **115,** 706–708.

29. Collins, C. B., Farquhar, R. M., and Russell, R. D. (1954). Isotopic Constitution of Radiogenic Leads and the Measurement of Geological Time. *Bull. Geol. Soc. Am.* **65,** 1–22.

30. Holmes, A. (1954). The Oldest Dated Minerals of the Rhodesian Shield. *Nature* **173,** 612–615.

31. Holmes, A., and Besairie, H. (1954). Sur quelques mesures de geochronologie à Madagascar. *Compt. rend.* **238,** 758–760.

32. Giletti, B. J., and Kulp, J. L. (1954). Radon Leakage from Radioactive Minerals. *Am. Mineral.* in press.

33. Collins, C. B., Farquhar, R. M., and Russell, R. D. (1952). Variations in the Relative Abundances of the Isotopes of Common Lead. *Phys. Rev.* **88,** 1275.

33a. Vinogradov, A. P., Zadurozhnyt, I. K., and Zykov, S. I. (1952). Isotopic Composition of Leads and the Age of the Earth. *Doklady Akad. Nauk S.S.S.R.* **85,** 1107–10.

33b. Kulp, J. L., and Bate, G. L. (1955). Isotopic Composition of Some Samples of Common Lead. *Phys. Rev.* in press.

34. Phair, G., and Levine, H. (1952). Notes on the Differential Leaching of Uranium, Radium, and Lead from Pitchblende. *Bull. Geol. Soc. Am.* **63,** 1290.

35. Starik, I. E. (1937). Questions of Geochemistry of Uranium and Radium, *Acad. Sci. USSR* Volumes dedicated to V. I. Vernadsky on 50th anniversary of his scientific activity, edited by Fersman.

36. Fleming, E. H., Jr., Ghiorso, A., and Cunningham, B. B. (1952). The Specific Alpha Activities and Half-lives of U^{234}, U^{235}, and U^{236}. *Phys. Rev.* **88,** 642–652.

37. Kulp, J. L., Broecker, W. S., and Eckelmann, W. R. (1952). Age Determinations of Uranium Minerals by Pb^{210}, *Nucleonics* **11,** 19–21.

38. Anderson, E. C., Libby, W. F., Wernhouse, S., Reid, A. F., Kirshenbaum, A. D., and Grosse, A. V. (1947). Radiocarbon from Cosmic Radiation. *Science* **105,** 576.

39. Libby, W. F. (1952). Radiocarbon Dating. University of Chicago Press, Chicago.

40. Arnold, J. R., and Libby, W. F. (1951). Radiocarbon Dates. *Science* **113,** 111.
41. Libby, W. F. (1951). Radiocarbon Dates. II. *Science* **114,** 291–296.
42. Libby, W. F. (1953). Chicago Radiocarbon Dates. III. *Science* **116,** 673.
43. Kulp, J. L., Feely, H. W., and Tryon, L. E. (1951). Lamont Natural Radiocarbon Measurements. I. *Science* **114,** 565–568.
44. Kulp, J. L., Tryon, L. E., Eckelmann, W. R., and Snell, W. A. (1952). Lamont Natural Radiocarbon Measurements II. *Science* **116,** 409–414.
45. Blau, M., Deevey, E. S., Jr., Gross, M. S. (1953). Yale Natural Radiocarbon Measurements I. Pyramid Valley, New Zealand and Its Problem. *Science* **118,** 1–6.
46. Carr, D. R., and Kulp, J. L. (1954). Dating with Natural Radioactive Carbon. *Trans. N.Y. Acad. Sci.* [II] **16,** 175–182.
47. Kulp, J. L., and Tryon, L. E. (1952). Extension of the Carbon-14 Age Method. *Rev. Sci. Instrum.* **23,** 296–297.
48. Smith, P. V., Jr. (1952). The Occurrence of Hydrocarbons in Recent Sediments from the Gulf of Mexico. *Science* **116,** 437–9.
49. Goldschmidt, V. M. (1937). Geochemische Verteilungsgesetze der Elemente IX. Die Mengenverhältnisse der Elemente und der Atom-arten. *Skrifter Norske Videnskaps-Akad. Oslo I Mat. Natur. Kl.* **4,** 140.
50. Ahrens, L. H. (1949). Measuring Geologic Time by the Strontium Method, *Bull. Geol. Soc. Am.* **60,** 217–266.
51. Davis, G. L., and Aldrich, L. T. (1953). Determination of the Age of Lepidolites by the Method of Isotope Dilution. *Bull. Geol. Soc. Am.* **64,** 379–380.
52. Ahrens, L. H., and Evans, R. D. (1948). The Radioactive Decay Constants of K^{40} as Determined from the Accumulation of Ca^{40} in Ancient Minerals. *Phys. Rev.* **74,** 279–286.
53. Aldrich, L. T., and Nier, A. O. (1948). Argon 40 in Potassium Minerals. *Phys. Rev.* **74,** 876–7.
54. Inghram, M. G., Brown, H. S., Patterson, C., and Hess, D. (1950). The Branching Ratio of K^{40} Radioactive Decay. *Phys. Rev.* **80,** 916.
55. Russell, R. D., Shillibeer, H. A., Farquhar, R. M., and Mausuf, A. K. (1953). Branching Ratio of K^{40}. *Phys. Rev.* **91,** 1223.
56. Shillibeer, H. A., Russell, R. D., Farquhar, R. M., and Jones, E. A. W. (1954). Radiogenic Argon Measurements. *Phys. Rev.* **94,** 1793.
57. Wasserburg, G. J., and Hayden, R. J. (1954). The Branching Ratio of K^{40}. *Phys. Rev.* **93,** 645.
58. Keevil, N. B. (1941). Helium Retentivities of Minerals. *Trans. Am. Geophys. Un.*, Part II, 501–3.
59. Keevil, N. B. (1941). The Unreliability of the Helium Index in Geological Correlation. *Univ. Toronto Studies Geol. Ser.* **46,** 39–67.
60. Hurley, P. M. (1950). Distribution of Radioactivity in Granites and Possible Relation to Helium Age Measurements. *Bull. Geol. Soc. Am.* **61,** 1–8.
61. Petterson, H. (1937). Lecture to the Vienna Academy of Science.
62. Holland, H. D., and Kulp, J. L. (1954). The Transport and Deposition of Uranium, Ionium, and Radium in Rivers, Oceans, and Ocean Sediments. *Geochim. et Cosmochim. Acta* **4,** 197.
63. Holland, H. D., and Kulp, J. L. (1954). The Mechanics of Removal of Ionium and Radium from the Oceans. *Geochim. et Cosmochim. Acta* **5,** 214–224.
64. Volchok, H. L., and Kulp, J. L. (1955). The Ionium Method of Age Determinations. *Geochim. et Cosmochim. Acta* to be published.
65. Ericson, D. B., Ewing, M., and Heezen, B. C. (1952). Turbidity Currents and Sediments in the North Atlantic. *Bull. Am. Assoc. Petrol. Geol.* **36,** 489–511.

66. Holland, H. D., and Kulp, J. L. (1950). Geologic Age from Metamict Minerals. *Science* **111,** 312.
67. Kulp, J. L., Volchok, H. L., and Holland, H. D. (1952). Age from Metamict Minerals. *Am. Mineral.* **37,** 709–718.
68. Hurley, P. M., and Fairbairn, H. W. (1953). Radiation Damage in Zircon; a Possible Age Method. *Bull. Geol. Soc. Am.* **64,** 659–673.
69. Daniels, F., and Saunders, D. F. (1951). The Thermoluminescence of Crystals, Final Report A.E.C. Contract AT(11-1)-27, Madison, Wisconsin.
70. Patterson, C. C., Goldberg, E. D., and Inghram, M. G. (1953). Isotopic Composition of Quaternary Clays. *Bull. Geol. Soc. Am.* **67,** 1387–1388.
71. Farquhar, R. M., Collins, C. B., Russell, R. D. (1954). Isotopic Constitution of Radiogenic Leads and Measurement of Geologic Time, *Bull. Geol. Soc. Am.* **65,** 1–27.
72. Kaufman, S., and Libby, W. F. (1954). The Natural Distribution of Tritium. *Phys. Rev.* **93,** 1337–1344.
73. Macnamara, J., and Thode, H. G. (1950). The Isotopes of Xenon and Krypton in Pitchblend and the Spontaneous Fission of U^{238}. *Phys. Rev.* **80,** 471.
74. Suttle, A. B., and Libby, W. F. (1954). Natural Radioactivity of Rhenium. *Phys. Rev.* **95,** 866.

Earthquake Seismographs and Associated Instruments

HUGO BENIOFF

Professor of Seismology, California Institute of Technology, Pasadena, California

1. Torsion Seismographs

1.1. Original Torsion Seismographs—Anderson and Wood

Progress in seismograph instrument development in the last quarter century has been quite rapid. One of the first of the new instruments was the torsion seismograph invented by J. A. Anderson and developed jointly by him and H. O. Wood [1]. In this instrument the inertia reactor is made up of a small slug of metal (usually copper) in the shape of a circular cylinder or a flat plate or vane, I (Fig. 1). The reactor is attached eccentrically to a taut wire or ribbon suspension S. Vibratory movement of the ground results in rotation of the inertia reactor about the suspension, and this movement is magnified and recorded optically by means of a mirror M_1, attached to the reactor. Electromagnetic damping of the eddy-current type is accomplished by immersion of the reactor in a magnetic field provided by a permanent magnet, F. The original models were constructed in two forms having periods of 0.8 seconds and 6 seconds respectively, the former for recording local earthquakes and the latter for teleseisms. The 0.8-second instrument is constructed with a copper cylindrical reactor 2 millimeters in diameter and 2.5 centimeters long. It weighs (with the mirror) approximately 0.7 gram. In order to double the magnification available with a given optical lever, a fixed mirror, M_2, is mounted near the moving system in such a manner that the recording beam is reflected twice from the moving mirror as shown in Fig. 1.

If the diameter of the cylindrical inertia reactor is d, the distance of

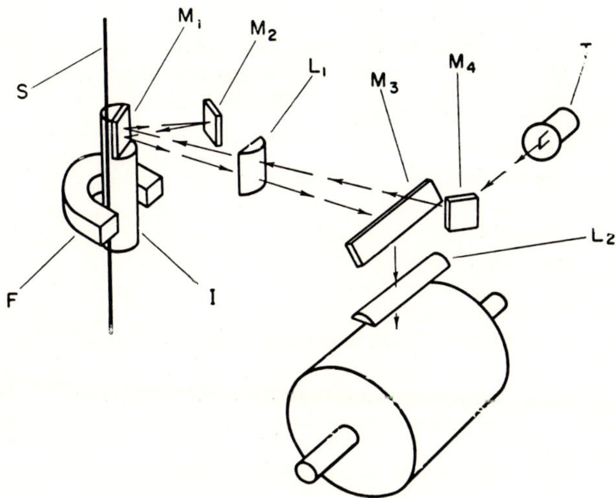

FIG. 1. Torsion seismograph (schematic).

the center of oscillation from the rotational axis is $l = \frac{3d}{4}$, and the magnification V is

$$V = \frac{4L}{l} \tag{1}$$

when a fixed deflection-doubling mirror is used. L is the optical lever arm—the distance from the seismometer lens to the recording drum. When $d = 2$ mm and $L = 1$ meter, $V = 2666$, the approximate value used in the original instruments.

In the vane-type instrument the distance of the center of oscillation from the rotational axis is

$$l = \frac{2d}{3} \tag{2}$$

In this case d is the width of the vane normal to the direction of the suspension. In the original vane-type instrument the period T_0 is adjusted to 6 seconds. The dimension of the vane is 7.5×25 mm. Using a double reflection from the moving mirror, as in the 0.8-second instrument, and a 1-meter optical lever, the magnification is

$$V = \frac{4L}{l} = \frac{4000}{\frac{2}{3} \times 7.5} = 800 \tag{3}$$

Higher magnifications are not practicable with direct-coupled horizontal-component pendulum instruments because of disturbances resulting from tilt of the ground.

In order to reduce transverse vibrations, Anderson passed the upper and lower suspensions through two perforated lugs filled with oil. In addition to reducing the transverse vibration, the oil dampers to some extent prevent horizontal displacement of the rotational axis during acceleration of the ground, and thus effectively improve the response.

The original design of the short-period torsion seismograph is subject to several difficulties. The mirror is mounted eccentrically relative to the inertia reactor, and consequently the directional response pattern is displaced angularly relative to the position defined by the inertia reactor alone. Moreover, as a result of unavoidable magnetic impurities in the material of the inertia reactor and the nonradial character of the magnetic field, the moving system is subject to a magnetic restoring force of unknown value. The free period as measured with the field removed is thus altered an unknown amount when the field is applied, and consequently in the operating condition the period is not known. A further disadvantage in this type of instrument is that a vertical component has not as yet been constructed, except in the very low magnification, strong motion types. This limitation is principally related to the thermal coefficient of elasticity of the suspension which must support the reactor against the acceleration of gravity.

In spite of the disadvantages mentioned above, the 0.8-second period torsion seismograph with static magnification of 2700 has proved to be a stable and useful instrument for the recording of the horizontal component of the stronger local earthquakes. Its limited magnification and the lack of a vertical component have seriously restricted its applicability. Likewise, the 6-second period instrument with magnification of 800 has proved generally useful in the recording of some of the waves of large shallow earthquakes. In the opinion of the writer the teleseismic form would be very much more effective if, instead of having a 6-second period, it were constructed with a longer period and smaller magnification, approximately the same as the corresponding constants of the Milne-Shaw instrument—the magnification being determined by the maximum tolerable tilt sensitivity. The shorter 6-second period was chosen as a result of an erroneous conclusion concerning the transient behavior of displacement-type seismometers, which in turn was a result of an incorrect assumption as to initial conditions in the solution of the differential equation of the seismometer as given by Anderson and Wood [1].

1.2. Strong-Motion Seismograph—Benioff

Following the Long Beach earthquake of 1933 the writer [2] designed a torsion seismometer having a magnification of 30 for recording aftershocks of that earthquake. The inertia reactor was made in the form of a rectangle of heavy copper wire mounted with the suspension concentric with one of the long legs of the rectangle. The magnetic field for damping was formed between the cylindrical surface of a central iron core pole-piece and an outer concentric cylindrical pole-piece. The outer leg of the inertia reactor thus moved tangentially in this field and the pendulum was therefore not subject to period changes by reason of magnetic forces arising from impurities in the copper rectangle.

1.3. Strong-Motion Seismograph—Lehner

In place of a single-turn coil reactor, one may substitute a coil of many turns with the terminals connected respectively to the upper and lower suspensions, which must be insulated from each other. Damping can then be controlled by varying the resistance in the external circuit connected to the suspensions. An instrument of this type with low magnification (100) was recently made by F. Lehner.

1.4. Strong-Motion Seismograph—Smith

A strong-motion torsion seismograph having a period of 10 seconds and a magnification of 4, when operating with a one meter optical lever, was designed and built by Sinclair Smith [3]. In this instrument the inertia reactor consisted of two unequal masses mounted on the ends of a short rod suspended at its center by a torsion ribbon. In order to prevent transverse translation of the ribbon during ground motion, the ribbon was stretched across two pivots positioned above and below the transverse arm of the inertia reactor. Two instruments of this type have been operating continuously at the Seismological Laboratory in Pasadena for many years.

1.5. Strong-Motion Seismograph—Wenner

Another strong-motion seismometer was designed by Frank Wenner [4]. This instrument has been used by the U. S. Coast and Geodetic Survey in a strong-motion program of earthquake investigation in California. It consists of a quadrifilar torsion suspension with a pendulum mass of 4 grams and a reduced length of 0.935 centimeters. The instrument was designed for use as an accelerometer for nearby earthquakes and has a period of 0.1 second with critical damping of the eddy-current type. When recording at a distance of 50 centimeters from the drum, a trace amplitude of 1 centimeter corresponds to a ground acceleration of 0.05 gravity.

1.6. Strong-Motion Seismograph—McComb

The Wenner accelerometer was later modified by H. E. McComb [5], who substituted a slender shaft with a pivot support for the quadrifilar suspension. Restoring force is supplied by a helical spring. For operation on the strong motion program of the U. S. Coast and Geodetic Survey, the recorder for this instrument was normally at rest. An electric contact starter designed by McComb [5] has the form of a 1-second pendulum carrying a platinum cone separated by a small distance from a fixed concentric platinum ring insulated from the pendulum. Small movements of the pendulum, such as those produced by artificial disturbances, do not produce contact of the pendulum. With large motions, such as those produced by strong local earthquakes, the cone contacts the ring and thus closes a relay circuit for setting the recorder into operation. The delay in starting is about 0.2 second.

2. Pendulum Seismographs with Viscous-Coupled Optical Recording Elements

2.1. Original Design by Romberg

Direct-recording pendulum seismographs, having periods greater than about 4 seconds, are severely limited as to their maximum useful magnification because of their high sensitivity to earth tilts in the case of the horizontal-component instruments, and to thermal response of the spring in the case of the vertical-component instruments. It is possible to eliminate these long-period drifts by means of a viscous coupling between the pendulum and the optical or mechanical recording elements. Such an instrument, first described by Arnold Romberg [6], is shown schematically in Fig. 2, in which R represents the inertia reactor of a conventional horizontal pendulum. The recording system includes a drum for photographic paper and a small mirror mounted on a vertical torsion suspension. Attached to the mirror is a short member, bent in the shape of an L, which dips into oil held in the small cup attached to the pendulum. The clearances between the vane and the cup and the viscosity of the oil are so chosen that, for relatively short-period movements of the pendulum, which are characteristic of earthquake waves, the coupling between the pendulum and the mirror is sufficiently tight to constrain the mirror to record the motion of the pendulum accurately. For the long-period movements characteristic of tilt, on the other hand, the coupling is weak enough to prevent an observable response.

2.2. Modified Design of McComb

A modification of the Romberg design, in which the restoring force of the mirror system is provided by gravity instead of by torsion of the sus-

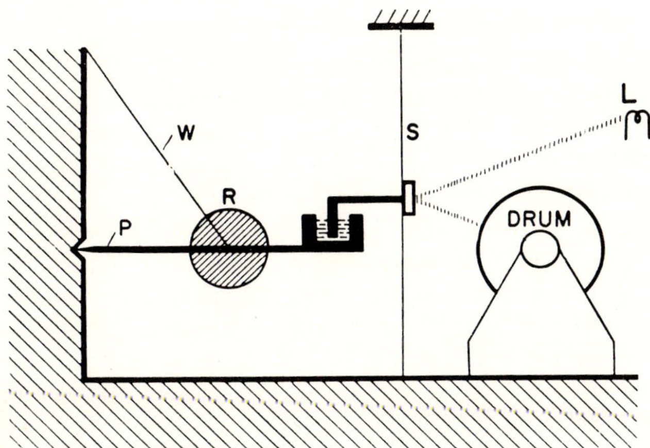

FIG. 2. Viscous coupled seismometer—Romberg.

pension, was later introduced by H. E. McComb [7]. In principle the modification is unimportant, provided constants are so chosen that the coupling element itself does not respond to vertical movements of the ground. So far, instruments of this type have been operated with low magnifications only (100 to 200).

3. Moving-Conductor Electromagnetic Pendulum Seismograph

3.1. Modified Galitzin Seismograph—Gutenberg

One of the first effective modifications of the moving-conductor transducer (Galitzin) seismograph was made by B. Gutenberg [8]. He shortened the period of the pendulum to 3 seconds (by tilting the base) and increased the magnetic field strength by decreasing the air gap clearance. This resulted in a frequency-response characteristic with higher magnifications for the short-period waves than the standard Galitzin or any other instrument in use at that time. The instrument demonstrated its superiority for short-period longitudinal waves in distant earthquakes as well as for recording of local shocks. The period-response characteristic (see Figs. 46 and 47) was such as to offer substantial magnification in the 6-second period range and the maximum useful magnification was therefore limited by microseisms of this period to approximately 4400.

3.2. Overdamped Galvanometer Seismograph—Wenner

Another substantial modification of the moving-conductor (Galitzin) electromagnetic seismograph was effected by Frank Wenner [9]. His modification consisted essentially of substituting an overdamped galvanometer for the critically damped galvanometer used by Galitzin. It can be shown that with overdamping, the response of a galvanometer is proportional to the time integral of the applied emf over an effective frequency band width. Since the emf generated by the transducer is proportional to the time rate of the pendulum displacement, the response of an overdamped galvanometer connected to the pendulum transducer is proportional to the pendulum displacement over a band of frequencies determined by the period of the galvanometer and the degree of overdamping. Wenner used a seismometer period of 12.5 seconds and a galvanometer with the same period. The Wenner seismograph made a satisfactory teleseismic instrument which in some respects, particularly its response at the shorter periods, was superior to the Galitzin design. He did not design a vertical-component instrument, and consequently this type of seismograph did not come into general use. There is no reason to believe, however, that a satisfactory vertical component can not be designed.

3.3. Galitzin Seismograph—Sprengnether

In recent years Sprengnether [10] has manufactured a Galitizin-type instrument which differs from the original only in mechanical details, and another one in which the constants are substantially identical with those of the Gutenberg modification.

3.4. Moving-Conductor Seismograph with Pendulum and Galvanometer Periods Unequal—Benioff

The newer magnetic alloys having high energy content, such as Alnico, have made it possible to design moving-conductor electromagnetic pendulum instruments with short periods and with response characteristics similar to those of the variable reluctance instrument described in a later paragraph. Figure 3 is a photograph of a vertical-component instrument of this type designed by the writer. The pendulum consists of a cylindric mass of 50 kg, attached to the frame by means of Cardan hinges. In this particular model, the spring is made up of a group of flat springs stressed in compression. The spring tension is communicated to the inertia reactor by means of a ribbon in such a way that the point of contact is defined by an adjustable bridging member for control of the period. The coil is mounted on the end of an aluminum tube and moves in an air gap having a field strength of about 12,000 gauss. By virtue of the lever action of the system, the coil movement is approximately 5 times that of the center of oscillation of the inertia reactor, thus reducing the required size of the magnet by a factor of 25 as compared with the system in which the ratio of coil displacement to pendulum displacement is unity. In normal operation the period of this instrument is adjusted to 1.5 seconds and the period-response characteristic is very nearly identical with that of the variable reluctance seismograph. The 50 kg pendulum mass, however, limits the recorder to one galvanometer at high magnification. The absence of negative restoring force in the magnetic structure requires the mechanical period to be the same as the working period, and the ruggedness of the instrument is therefore less than that of the variable reluctance seismograph. It does have the advantage, however, of negligible reactance in the transducer coil. Figure 4 shows a photograph of the horizontal-component instrument of this type.

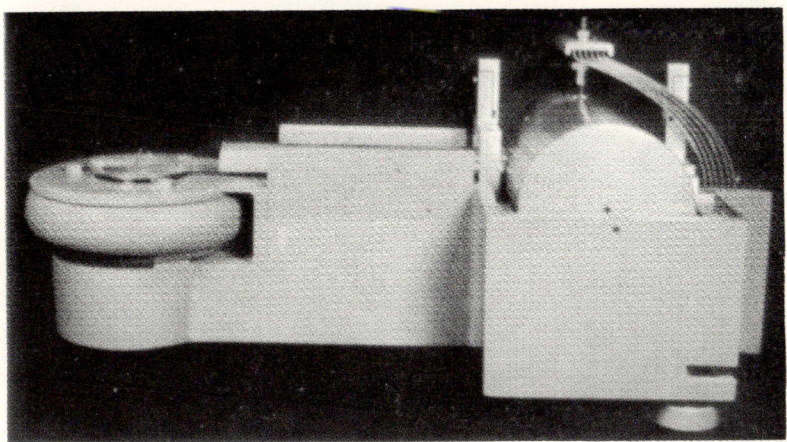

FIG. 3. Short-period vertical-component moving-conductor seismometer—Benioff. The cover is removed.

It is essentially similar to the vertical component, except that restoring force is provided by gravity, rather than by a spring. Typical response characteristics of these seismographs are shown in Figs. 46 and 47.

3.5. Long-Period Seismograph—Ewing and Press

Seismograms written by the Pasadena variable reluctance vertical-component seismograph, 115-seconds period galvanometer, and the long-period strain seismograph (T_g = 180) indicated the existence of Rayleigh waves having periods as long as 8 minutes. In order further to increase the sensitivity for these waves, Maurice Ewing and Frank Press [11] developed a vertical-component moving-conductor electromagnetic seismograph having a pendulum period of 15 seconds and a galvanometer period of 75 seconds. The pendulum is supported by a LaCoste spring and is otherwise conventional. Horizontal-component instruments having

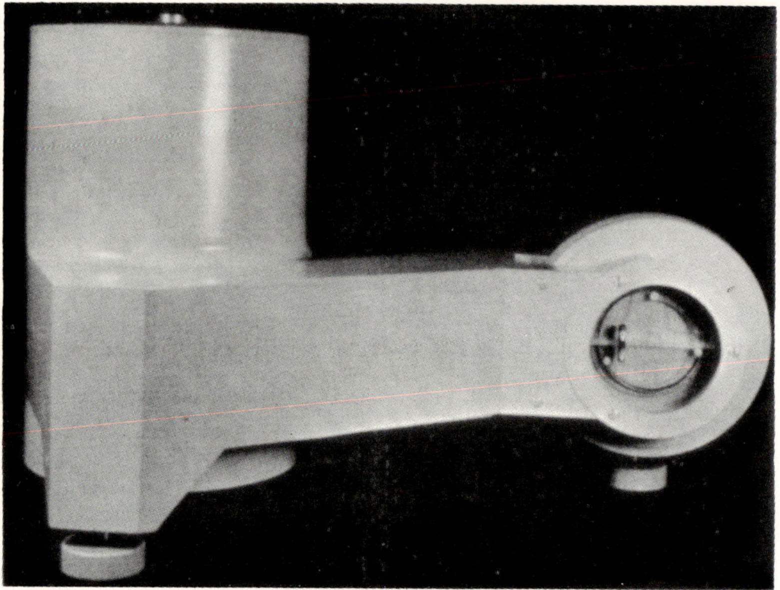

Fig. 4. Short-period horizontal-component moving-conductor seismometer—Benioff.

the same characteristics have also been constructed. Typical response curves for these seismographs are shown in Figs. 46 and 47.

4. Variable Reluctance Electromagnetic Pendulum Seismograph

4.1. Variable Reluctance Transducer—Benioff

The response of the Galitzin seismograph, even with Gutenberg's modification, was limited in the magnification available for short periods. For constant power in seismic waves the amplitude varies inversely with the period. Moreover, for periods less than approximately 4 seconds the microseismic activity decreases sharply with period. Thus with an electromagnetic seismograph, to record effectively the period range down to approximately $\frac{1}{5}$ second with magnification adequate to reach the microseismic level, it is necessary that the pendulum mass be large, say 50 to 100 kg, and that it have a period in the neighborhood of 1 second. Until recently, the magnetic materials available were such that it was impracticable to provide magnetic field strengths sufficient to damp such a pendulum by the reaction of the output currents, a condition which is necessary in order to derive the maximum power from the pendulum. This limitation has been overcome by the development of the variable reluctance transducer [12]. In effect it represents a modification of the ordinary telephone receiver, in which a permanent magnet supplies flux across two or more air gaps to an armature around which is wound a coil of wire. A portion of the transducer is attached to the frame of the instrument, while the rest is attached to the moving pendulum in such a way that movement of the pendulum relative to the frame of the instrument changes the lengths of the air gaps and thus the reluctance of the magnetic circuit. This, in turn, changes the magnetic flux through the armatures, and so generates emfs in the coils surrounding the armatures.

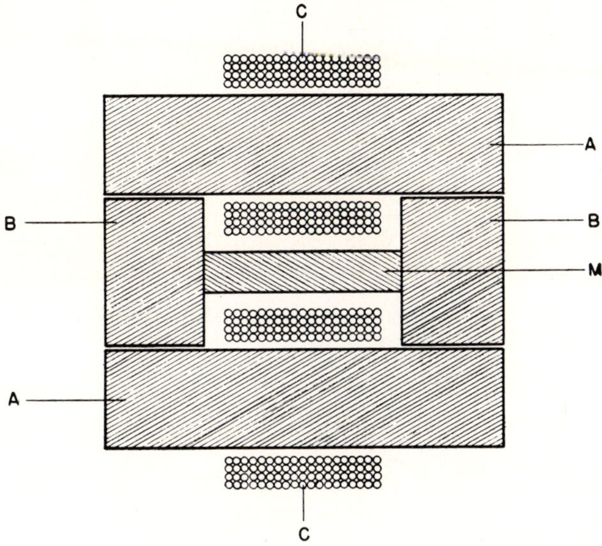

FIG. 5. Variable reluctance transducer—Benioff (schematic section).

In the original form the transducer was single-ended; that is, it contained one pair of air gaps, and as such was of relatively low efficiency, so that additional damping for the pendulum had to be provided by a dashpot mechanism with oil as the damping medium. The original single-ended transducer was modified [13] to the push-pull form, shown schematically in Fig. 5, which represents a section through the transducer. The magnet M, in the form of a square plate 7.5 by 1.6 centimeters, is in contact at two of its ends with the flux distributing members B, which are formed of laminations. In the early models these were made of ordinary silicon transformer iron. At the present time they are being made with a high-permeability, low-hysteresis alloy. Flux from the magnet is thus transmitted by the members B across the four air gaps to the two armatures A, which are bolted to brass plates to form a single moving system. Coils of wire C are wound around the armatures A. The armature portion of the transducer is attached to the pendulum, and the portion comprising the magnet M and the distributing members B is attached to the base of the instrument (see Fig. 6). When the pendulum is in its rest position, the four gaps are equal and are approximately 2 millimeters in length. Movement of the pendulum relative to the frame increases the lengths

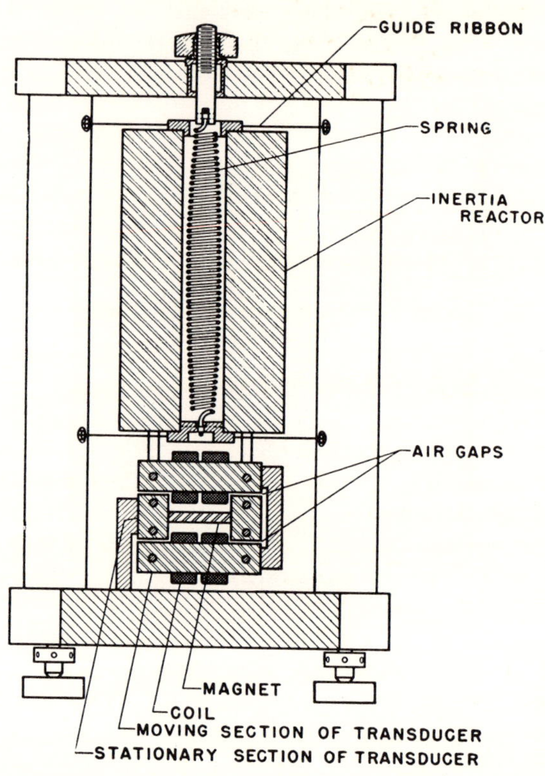

FIG. 6. Vertical-component variable reluctance seismometer—Benioff (schematic section).

of one pair of gaps and decreases the length of the other pair. The change in flux in the armatures A, resulting from the corresponding change in the reluctance of the air gaps produce emfs in the coils which are connected together in series or parallel in such a way that their separate emfs are additive. By virtue of the push-pull character of the transducer, the output emf is a linear function of the velocity of the pendulum up to terms of the third order in the pendulum displacement measured relative to the air gap length. The effectiveness of this instrument is such that, with the small magnet described, a pendulum with a 1-second period and mass of 100 kg can be damped critically by the reaction of the output currents when the transducer works into a load having a resistance equal to that of the transducer winding. In addition to its greater effective use of magnets, this transducer has an additional advantage over the moving-conductor transducer, which arises from the negative restoring-force characteristic of the magnetic traction across the air gaps. It was found experimentally that in order to attain critical damping the negative magnetic restoring force of the transducer must equal or exceed approximately $\frac{9}{10}$ the positive mechanical restoring force of the springs. Thus for an operating period of 1 second the mechanical period of the pendulum without the magnetic field is approximately 0.31 seconds. A seismometer having this transducer can therefore be approximately 9 times as rugged for the same period as one not employing negative restoring force. Moreover, since the air gaps are large (2 millimeters), the instrument does not require close tolerances in manufacture or adjustment.

4.2. Variable Reluctance Vertical-Component Seismograph—Benioff

A simplified schematic section of the vertical-component variable reluctance seismometer is shown in Fig. 6. The inertia reactor of 100 kg is in the form of a hollow cylinder of steel. It is supported by a nickel alloy spring, having a low thermal coefficient of elasticity. The pendulum is constrained to move in a vertical line by the six radial guide ribbons. These ribbons are attached to the three upright columns of the frame by means of flat springs (not shown in this diagram) which serve to maintain constant tension on the springs during motion of the pendulum. An adjustment on the flat springs alters the tension of the ribbons and serves to vary the period of the pendulum. The transducer is provided with eight coils of 125 ohms resistance each. The output power is sufficient to operate two galvanometers simultaneously. The standard assembly consists of one galvanometer with a period of 0.2 seconds and a second galvanometer of 90- to 100-seconds period. The short-period galvanometer has a resistance of 25 ohms and is driven by four of the coils of the transducer connected in parallel, two from one armature and two from the other. By using coils from each armature in equal numbers, maximum linearity of output is obtained and, in addition, the response to stray electromagnetic fields is greatly reduced. The maximum available magnification for the short-period combination is of the order of 200,000, sufficient to show the ground unrest at any site on the earth. The long-period galvanometer usually has internal and critical damping resistances of 500 ohms and is operated by the remaining four coils connected in series. In the original model, in which the transducer laminations were of silicon iron, damping of the pendulum motion resulting from hysteresis losses in the laminations was rather large. With the alloy laminations this effect is negligibly small. The inductive reactance of the windings of the transducer is appreciable for periods in the vicinity of the pendulum period and shorter. Consequently, for these periods the currents in the coil lag somewhat behind the pendulum velocity and the generated emf, with the result that the damping cannot be strictly critical, although the departure is small. Furthermore, for a certain period, which is very much shorter

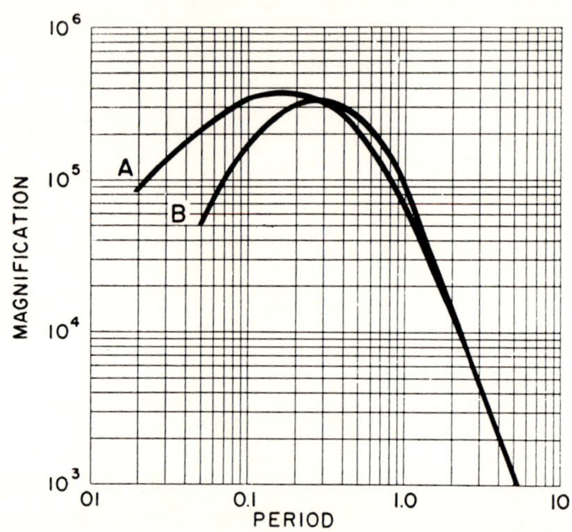

FIG. 7. Period-response characteristic of short-period variable reluctance seismograph. $T_0 = 1$ sec, $T_g = 0.189$ sec, $h_0 = 1$, $h_g = 1$. Curve A is calculated assuming no inductive reactance and no interaction from the equation $Q = \dfrac{T}{2\pi(U^2 + 1)(V^2 + 1)}$, $U = T/T_0$, $V = T/T_g$, T = ground period, T_0 = pendulum period, and T_g = galvanometer period. Curve B was measured with a shaking table by Jack Hamilton of the Geotechnical Corporation.

than that of the pendulum, the lag is such that the output currents provide a positive magnetic restoring force which, added to that of the spring, produces a second mode of oscillation of very short period and low damping. Usually this period is outside of the working range of the instrument. A further effect of the reactance of the windings is a reduction of magnification in the short-period range in relation to that which obtains with a moving-conductor electromagnetic seismograph. A measured response curve of the short-period galvanometer combination is shown in A, Fig. 7. This was obtained on a shaking table by Mr. Jack Hamilton of the Geotechnical Corporation, the present manufacturers of the instruments. The calculated response characteristics for an electromagnetic seismograph having the same constants, but with zero transducer inductance and with sufficiently large pendulum mass to be free of

FIG. 8. Vertical-component variable reluctance seismograph—Benioff—manufactured by Geotechnical Corporation.

galvanometer reaction, is shown at B. Figure 8 is a photograph of the vertical-component instrument with the outer covers removed. Its temperature stability is outstanding. The pendulum remains stable and in operating condition over a temperature range of approximately 55° C without adjustment. Figure 9 is a copy of a portion of the seismogram of the great Assam earthquake of August 15, 1950, written with the 115-

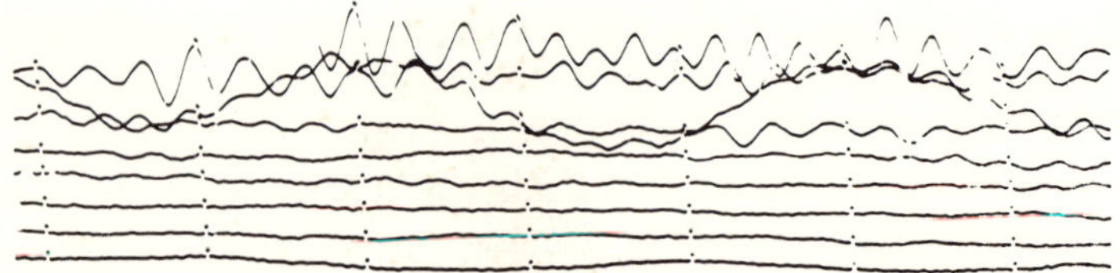

FIG. 9. Portion of the seismogram of the Assam earthquake of August 15, 1950, written at Pasadena with a vertical-component variable reluctance seismograph with a galvanometer of 115-seconds period. Note Rayleigh waves with approximately 3-minutes period.

second galvanometer variable reluctance seismograph, showing Rayleigh waves of approximately 3-minutes period.

4.3. *Variable Reluctance Horizontal-Component Seismograph—Benioff*

In the horizontal-component seismometer the steady mass is divided into two sections rigidly attached to the transducer mounted between them. The reactor is supported by two of the six guide ribbons and is otherwise similar to the vertical-component instrument. Restoring force is provided by gravity and by tension of the ribbons. Figure 10 is a photograph of an older model of the instrument.

FIG. 10. Horizontal-component variable reluctance seismometer—Benioff. Early model.

5. Electrostatic Transducer Pendulum Seismographs

5.1. *Electrostatic Seismograph—Benioff*

The electrostatic transducer seismometer is modeled after the condenser microphone of acoustics. In one form [14], the inertia reactor (A, Fig. 11) consists of a circular brass plate 4 inches in diameter suspended by a single flat spring S and insulator I to form a pendulum with a period of 0.2 seconds. A second similar plate E is mounted to the frame of the instrument, parallel to the upper plate and separated from it by approximately 0.2 mm. Critical damping is provided by the viscosity of

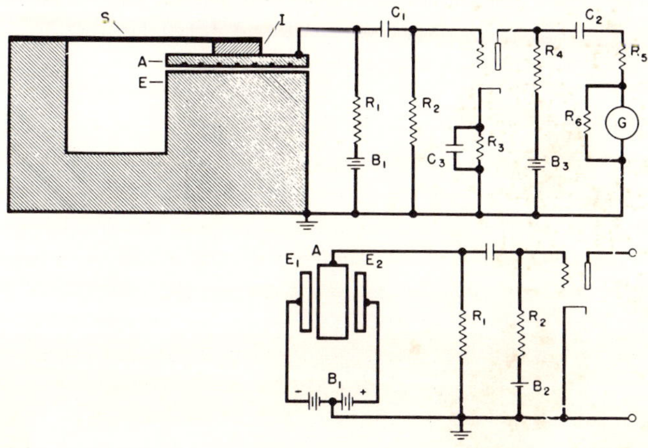

Fig. 11. Schematic representation and circuit diagram of electrostatic seismometer—Benioff. The upper diagram is of a single-ended transducer; the lower circuit is a recommended push-pull arrangement.

the air between the plates. Two sets of linear grooves milled at right angles to each other in the lower surface of the upper plate serve to control the amount of damping and also to reduce the elastic and inertia effects of the air between the plates. The insulated plate A is connected to one terminal of a battery B_1 through the high resistance R_1. The other terminal of the battery is connected to the frame of the instrument. Movement of the ground results in changes in the separation of the two plates, which thus produce changes in the capacity between them. The moving plate is coupled to the grid circuit of an amplifier tube by means of the coupling condenser C_1 having a capacity which is large in comparison with the seismometer capacity. The output of the tube is coupled through R_4 and C_2 to a short-period recording galvanometer. The galvanometer circuit impedance is matched to the tube output impedance by R_5. The critical damping resistance of the galvanometer is R_6. For earth periods which are short in comparison with the time constants of the input and output circuits, the output emf is proportional to the pendulum displacement for small displacements. The sensitivity of the transducer (exclusive of the amplifier) is about 2500 volts per centimeter displacement of the plates.

Although the single-ended structure operates satisfactorily for small displacements, considerable improvement in linearity and stability is gained by the push-pull modification shown in the lower half of Fig. 11. In this form the pendulum A is positioned midway between two fixed plates E_1 and E_2, with the midpoint of B_1 grounded. The rest of the circuit is the same as for the single-ended transducer. The electrostatic transducer introduces a negative restoring force similar to that of the variable reluctance electromagnetic transducer.

5.2. Electrostatic Seismograph—Gane

Another form of electrostatic transducer seismograph was developed by P. G. Gane [15]. He uses a torsion pendulum with an inertia reactor having the form of a cylindrical tube of aluminum. Fixed electrodes of iron with concave cylindrical surfaces are positioned close to the inertia reactor to form two cylindrical air gaps of approximately 0.8 mm in length. The pendulum is damped partly by air damping but principally by eddy-current damping provided by flux from a permanent magnet having iron electrostatic electrodes and a central fixed iron core for pole-pieces. The period of the pendulum is 0.23 seconds. The two iron electrodes are insulated from each other and from the rest of the structure. One of the electrodes is connected to the ground and the other to a 150-volt source. The inertia reactor is also insulated from the rest of the structure and is connected to the grid of a cathode follower tube having no external grid leak. The capacity of the condenser is approximately 60×10^{-12} farads, and it is claimed that the effective input resistance of the tube, a metal 6J7, is 10^9 ohms giving a time constant of about 6×10^{-2} seconds. Gane's circuit connections form a single-ended transducer although the mechanical structure is such as to be easily wired for push-pull operation. Owing to the short pendulum period, a single design of the seismometer serves for either vertical or horizontal component response. It is clear that this instrument, as well as the one previously described, is suitable for recording very short-period earthquake waves only. The time constant of the electrostatic seismometer could be increased by a factor of 10 or 100 by using larger plates, but even under this condition it would be severely limited as to period response unless used with amplifiers having input resistances of the order of 10^{11} or 10^{12} ohms. Under these conditions leakage currents over the insulation, particularly those arising from dust particles between the plates, would be serious. For these reasons the electrostatic transducer instrument has not found use except in special applications involving short periods only.

6. Carrier-Current Transducer Seismographs

6.1. *Capacity Bridge Seismograph with Nonphase-Sensitive Detector—Volk and Robertson*

A seismometer employing a fixed carrier current for operation of a capacitance-bridge transducer was described by Volk and Robertson [16]. The pendulum of the seismometer carries a plate which is positioned

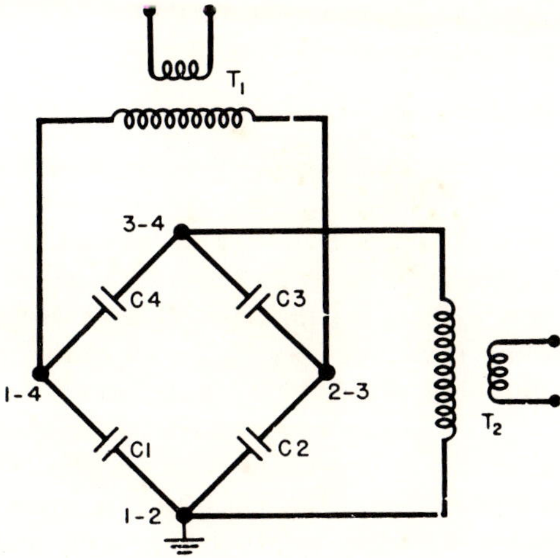

Fig. 12. Circuit diagram of a carrier-current capacity bridge transducer.

midway between two fixed plates. The capacitances between the pendulum plate and the fixed plates are represented by C_1 and C_2 in the bridge circuit of Fig. 12. C_3 and C_4 are fixed capacitors. The bridge is supplied with power at 8.6 megacycles, which is derived from a 50-watt crystal controlled oscillator, and transmitted at 100 ohms impedance through a cable to the transformer T_1 situated at the bridge. The point 3-4 of the bridge is connected directly to the grid of the first tube of a two stage rf amplifier rather than to the transformer T_2 shown in the figure. When the pendulum is in the equilibrium position the bridge is balanced, and the rf voltage supplied to the grid of the amplifier tube is zero. Movement of the pendulum unbalances the bridge and produces an rf voltage to the amplifier. The output of the amplifier is rectified by the diode section of a duodiode-triode tube and the dc output of the diode is impressed on the grid of the triode for dc amplification. The quiescent triode plate current is balanced out in a resistive bridge circuit, and the unbalance serves to operate the galvanometer or recorder. Actually this system does not function with the bridge normally in balance, since the detector is not phase sensitive. The quiescent position of the pendulum must therefore represent an unbalance of the bridge which is greater than any unbalance expected from seismic vibration or from thermal or tilt drift of the pendulum. Although high stability and magnifications up to a million are claimed, recordings demonstrating this behavior were not reproduced in the paper cited. Since the bridge must operate in a normally unbalanced position, the stability of the output must depend upon amplitude stability of the oscillator, which appears to be of the order of one percent. Consequently it would appear that the claims for this instrument are excessive.

6.2. Capacity Bridge Seismograph with Phase-Sensitive Detector—Cook

A carrier-current resonant bridge transducer employing a phase-sensitive detector is described by Cook [17] and applied by him to a vertical component seismometer of long period. He uses transformer input and output connections to the bridge as shown in Fig. 12. As constructed, this system was quite complex, and the reader is referred to the original article for details. No information is available as to its operation in the recording of earthquakes.

6.3. Variable Discriminator Transducer—Benioff

Another carrier-current seismograph having a variable discriminator transducer [18] is shown schematically in Fig. 13. In this form the capacitances C_1 and C_2, formed between the pendulum mass and two insulated

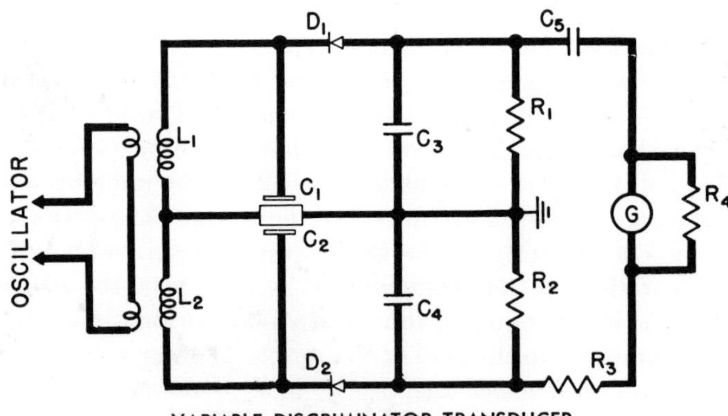

FIG. 13. Circuit diagram of a variable discriminator transducer—Benioff. G is the recording galvanometer, R_4 is the critical damping resistance of the galvanometer.

fixed plates, are shunted by two equal inductances, L_1 and L_2 respectively, to form resonant circuits. Current from a crystal-controlled rf oscillator is transmitted by a low impedance line loosely coupled to the inductances L_1 and L_2. In the equilibrium position of the pendulum the circuits, L_1C_1 and L_2C_2, are adjusted to the same resonant frequency either above or below that of the carrier by an amount which lowers the current to 0.7 of the resonant value as shown at O on the solid curve of Fig. 14. Thus in the quiescent condition the rf voltages across C_1 and C_2 are equal, and the rectified currents of the silicon diodes D_1 and D_2,

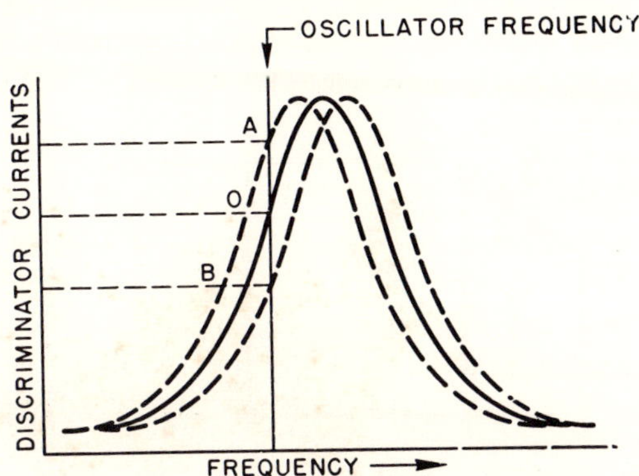

FIG. 14. Operational characteristics of a variable discriminator transducer. The solid curve is the current-resonant characteristic of each of the tuned circuits when the pendulum is in the quiescent position. The dashed curves show the two resonant characteristics as modified by a displacement of the pendulum.

flowing through R_1 and R_2, are thus equal and of opposite signs. Movement of the pendulum increases one capacitance and decreases the other. This results in shifts of the resonant curves of the two circuits, with a corresponding decrease and increase of the currents in the two as shown at A and B in the figure. The corresponding dc voltages across R_1 and R_2 are no longer equal, and a current thus flows through the output circuit including R_3, G (the recording galvanometer) and the large capacitor C_5. For applications requiring response to all frequencies down to zero, the condenser C_5 is omitted, but for regular use it is inserted in order to eliminate drift due to temperature effects on a vertical pendulum, and that due to tilt on a horizontal-component pendulum. If R_3 is 10 megohms and C_3 is 100 microfarads, the time constant of the galvanometer circuit is 1000 seconds. For best results C_3 is an oil-filled paper condenser, although for short-period waves electrolytic condensers have been found satisfactory. Since the power output of this system is large, the galvanometer can be insensitive and of short period. A Cambridge Instrument Company "Flik" galvanometer having a period of 0.1 second and external damping resistance of 7500 ohms has proved very satisfactory as a recorder for this instrument when operating with a pendulum having a period of 1.5 seconds. With transducers of this type the transducer output voltage is proportional to the pendulum displacement. Hence the period-response characteristic corresponds to that of the pendulum alone, for all periods longer than about twice the galvanometer period. The system has operated successfully with a crystal-controlled transistor oscillator and with a crystal-controlled tube oscillator. The transistor oscillator is operated at a frequency of 100 kilocycles. The circuit for this oscillator was adapted jointly by L. Blayney and the writer from Peter Sulzer [19], and is shown in Fig. 15. The oscillator and amplifier derive

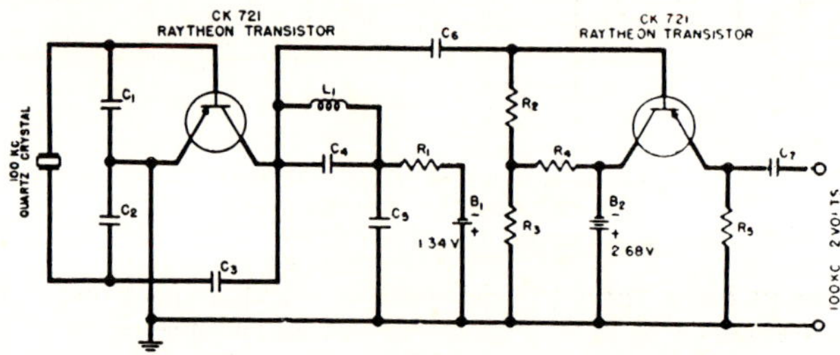

FIG. 15. A 100-kc crystal-controlled transistor oscillator for supplying carrier current to a variable discriminator seismograph.

their power respectively from 1.34-volt and 2.68-volt hearing-aid batteries of the mercury type. The current drain is so small that the operating life on one set of batteries is approximately one year. The output of the transistor amplifier is 2 volts and, when coupled to the discriminator, produces a dc output of 20 volts across each of the diode resistors R_1 and R_2. The crystal-controlled tube oscillator shown in Fig. 16 operates at 5.35 megacycles, and is of the two-tube type with the crystal X serving as the cathode coupling element between the two tubes VT, which are twin units in a single duotriode. The oscillator output is amplified by a buffer stage of duotriodes with cathode-follower outputs to two coaxial lines for operation of two seismometers (the discriminator associated with the second coaxial output line A in the figure is not shown). The quiescent frequency of the two resonant circuits of the discriminator is 5.40 megacycles. This discriminator differs from that shown in Fig. 13 in that one terminal of the galvanometer is at ground potential. This requires the diodes to be poled oppositely. The ganged switches SW_2 with resistors R_{12} to R_{26} serve to control the galvanometer current and hence the magnification. The galvanometer coupling condenser C_{12} eliminates drift. A pendulum stabilizing circuit is also shown. Here L_7 represents the damping coil of the seismometer, and R_9 is the critical damping resistance of the pendulum. When SW_1 is closed the dc unbalance current and the ac currents having frequencies too low to pass through C_{12} are thus fed into the seismometer damping coil with a polarity such that the

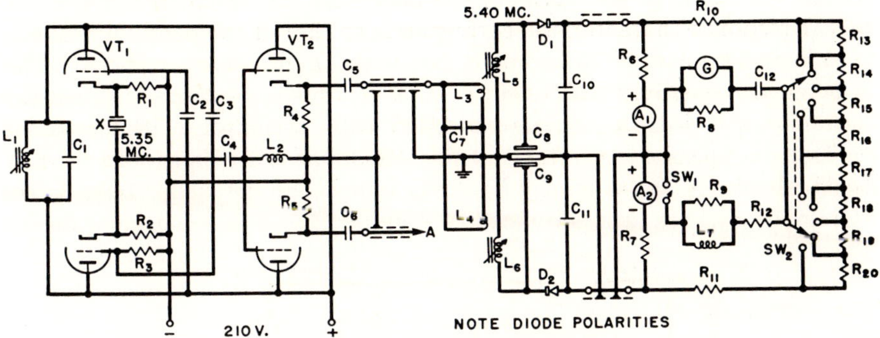

C_8, C_9 = SEISMOMETER CAPACITORS
L_7 = SEISMOMETER DAMPING COIL
A = TO SECOND SEISMOMETER

FIG. 16. Circuit for a variable discriminator seismograph using a vacuum tube oscillator at a frequency of 5.35 megacycles and galvanometric recording with one galvanometer terminal grounded—Benioff.

magnetic field which they produce reacts with the field of the damping magnet to oppose the pendulum drift. This modification reduces pendulum drift by a factor of approximately 0.25, and is useful in applications such as in portable installations where the pendulum drift may be large.

6.4. Horizontal-Component Variable Discriminator Seismograph—Benioff

A horizontal-component seismometer designed for operation with the variable discriminator transducer is shown schematically in Fig. 17. The

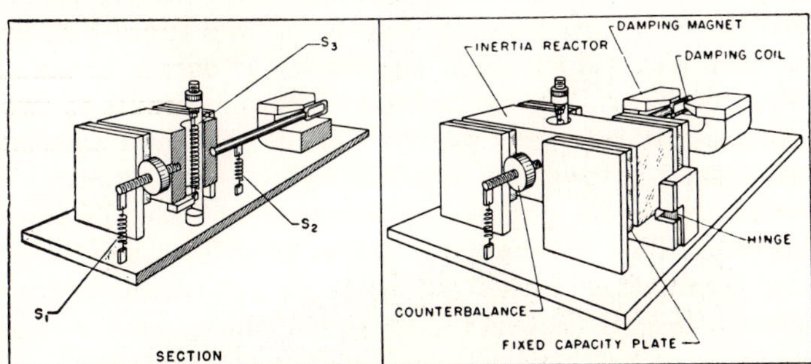

FIG. 17. Schematic drawing of horizontal-component variable discriminator seismometer—Benioff.

inertia reactor is a bar of brass 7.5 cm long with a square section 5 cm on the side. It is supported by two vertical tension hinges. Restoring force is provided partly by the stiffness of the hinges and partly by the helical springs S_1 and S_2. Negative restoring force is added to the system by gravity and by the spring S_3, which exerts a vertical tension at a point in the lower surface of the inertia reactor. The period is adjustable from about 0.5 second to 4 seconds, but normally is set at 1 to 1.5 seconds. A lever arm carries a small coil, one leg of which is immersed in a magnetic field for damping of the pendulum. The terminals of the damping coil are carried out through two helices made of fine copper wire to insulated terminals. In this way damping can be adjusted by varying the resistance shunted across the coil. A counterbalance is arranged opposite the damping coil lever to balance the pendulum for vertical movements of the ground. Four insulated plates are mounted as shown relative to the inertia reactor. They are separated from it by an air gap of 1.0 mm. The two plates on each side are connected together to form the two discriminator capacitors. The plates are grooved in order to reduce air damping effects.

6.5. Vertical-Component Variable Discriminator Seismograph—Benioff

Figure 18 shows a photograph of the vertical component seismometer which has the same size inertia reactor as the horizontal instrument. The capacitor plates are mounted above and below the reactor. The weight of the reactor is supported by a short negative-length helical alloy spring mounted within a hole bored in the inertia reactor. The spring is supported by a cantilever projecting over the inertia reactor. Typical

FIG. 18. Photograph of vertical-component variable discriminator seismometer—Benioff.

response curves for these seismographs with pendulum periods of 1.5 seconds are shown in Figs. 46 and 47.

6.6. *Variable Discriminator Circuits for Magnetic Tape and Visible Writing Recorder*

The variable discriminator seismograph has proven exceptionally satisfactory for operation of magnetic tape recorders and visible writing recorders of the ink and hot stylus types. Additional circuit elements required for simultaneous operation of a magnetic tape recorder and a visible writer are shown in Fig. 19. The portion of the circuit to the left, including resistors R_1 and R_2, represents the discriminator, and is identical

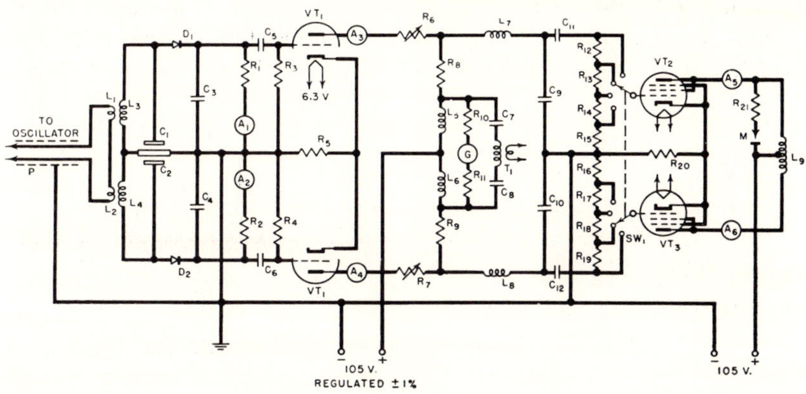

FIG. 19. Circuit of the variable discriminator seismograph, arranged for magnetic tape recording and visible writer recording.

with the corresponding part shown in Fig. 13. Condensers C_5, C_6 and resistors R_3, R_4 couple the discriminator output to the twin amplifier tube VT_1, operating in push-pull fashion. Resistors R_6 and R_7 serve to balance the gains of the two tubes. The two halves of a recording head winding are represented by L_5 and L_6. G is a monitoring galvanometer which, with R_{10} and R_{11}, indicates the seismic currents being recorded on the tape. The condensers C_7 and C_8 couple the output of transformer T_1 to the recording head for bias. The primary of the bias transformer operates from the 60-cycle power line. Since the tape recording speed is only 0.5 mm per second, the 60-cycle bias frequency becomes 45 kilocycles when the tape is played back at 15 inches per second for analysis. For tape recording only, without the visible writer, the portions of the circuit to the right of the recording head are omitted. L_7 and L_8 together with condensers C_9 and C_{10} form low-pass filters for preventing the 60-cycle bias current from exciting the grids of the writing amplifier tubes VT_2 and VT_3. The dual potentiometer SW_1 controls the input to the tubes

VT_2 and VT_3 and serves to adjust the sensitivity of the recorder. VT_2 and VT_3 are triode connected beam pentodes whose anodes are coupled directly to the split winding L_9 of the writing galvanometer coil. Time marks on the recording are introduced by unbalancing the currents in the two halves of the writing galvanometer winding by means of the relay contact M and resistance R_{21}. If the visible writer is operated without the tape recorder, L_5, L_6, L_7, L_8, and the associated recording head portions of the circuit are omitted.

7. Linear Strain Seismograph

7.1. *Original Strain Seismograph—Benioff*

Unlike pendulum seismographs, which respond to ground vibration, the strain seismograph [20] responds to strains of the ground. The principle of the strain seismograph is illustrated in Fig. 20, in which A and B

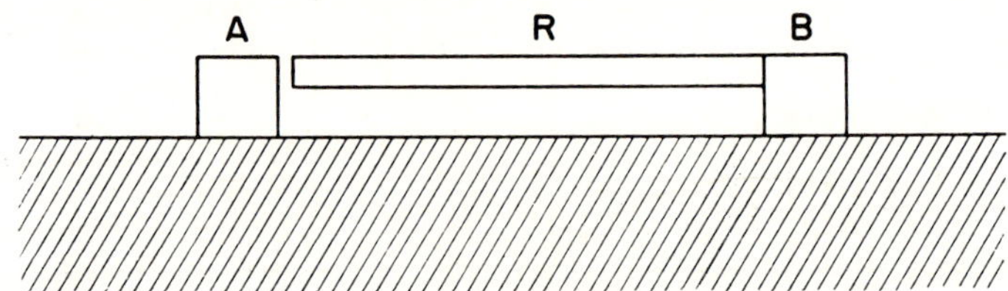

Fig. 20. Principle of the strain seismometer.

are two piers anchored to the rock and separated by distances of 20 to 50 meters. R is a rigid tube of steel or fused quartz attached to pier B and extending to within a short distance of pier A. The tube is supported by a group of members, such as shown in Figs. 23 and 27, designed for transverse rigidity and longitudinal flexibility. Strains in the ground, produced by seismic waves, change the relative positions of the two piers, and thus also the separation of the end of the tube R, with respect to the adjacent pier A. By interposing a suitable transducer between pier A and the end of the tube R, this change in separation, proportional to the ground strain, is magnified and recorded. In the original installation the tube was made of steel. In later models fused quartz was used. Since the surface of the earth is a node for vertical strain, the strain seismograph is effective primarily as a horizontal-component instrument. When body waves are incident at the earth's surface, they produce apparent waves traveling in the surface with speeds depending upon the angle of incidence as well as the body wave speed. The strain seismograph responds to these apparent waves, as well as to true surface waves. The response of the strain seismometer to longitudinal apparent or surface waves is

$$Y_L = -\frac{L}{C}\cos^2\alpha\,\frac{\partial\xi}{\partial t} \tag{4}$$

where L is the length of the indicator rod, C is the horizontal velocity of the wave, α is the angle between the direction of the rod and the direction of propagation of the wave, and ξ is the ground particle displacement. For apparent transverse waves or for the H-component of such waves the response is

$$Y_T = \frac{L}{C}\sin\alpha\cos\alpha\,\frac{\partial\xi}{\partial t} \tag{5}$$

Figure 21 shows the directional response characteristic of the strain seismometer for longitudinal waves. The pendulum characteristic is

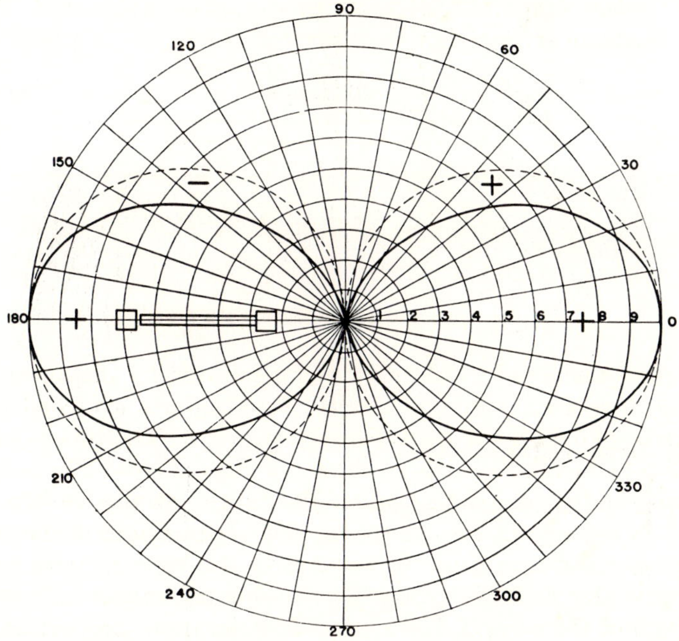

FIG. 21. Directional characteristic of strain seismometer for longitudinal waves.

shown in dotted lines. In Fig. 22 the directional characteristic of the linear strain seismograph for transverse apparent waves is shown.

If two strain seismometers are oriented at right angles with respect to each other, their respective responses to a given compressional wave are of like phase for all wave azimuths and of unequal amplitudes for all wave azimuths except those which bisect the angle between the two instruments. For transverse waves, however, the responses of the two instruments are equal in amplitude for all azimuths of the incoming wave, and are always opposite in phase. Comparison of phases on seismograms of the two components thus makes differentiation between shear waves and compressional waves easy. Since the shear wave responses of two perpendicularly oriented seismographs are always equal in amplitudes

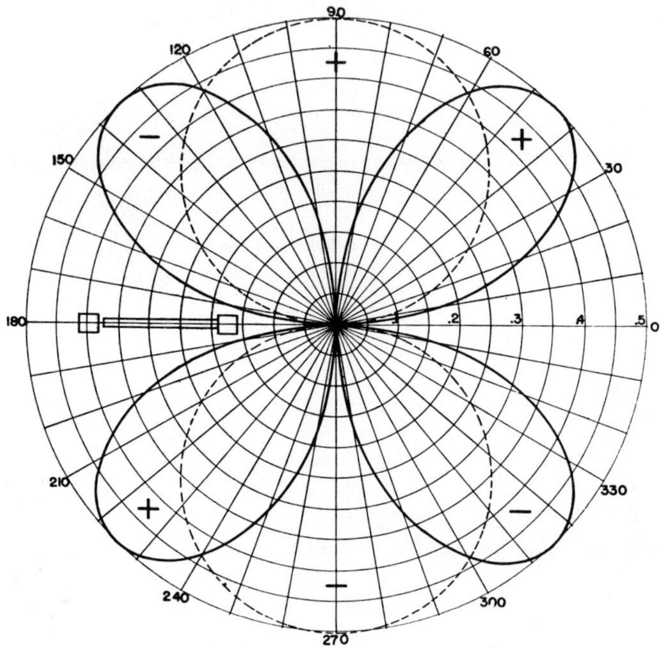

FIG. 22. Directional characteristic of strain seismometer for transverse waves.

and opposite in phase, a recording made by adding the responses of the two instruments will be zero for all transverse waves and for longitudinal waves it has the form

$$(6) \qquad Y_{L1} + Y_{L2} = -\frac{L}{C}\cos^2\alpha\,\frac{\partial\xi}{\partial t} - \frac{L}{C}\cos^2\left(\frac{\pi}{2} - \alpha\right)\frac{\partial\xi}{\partial t} = -\frac{L}{C}\frac{\partial\xi}{\partial t}$$

That is, the response to longitudinal waves will be the same for all directions.

Given a favorable site, the strain seismograph is inherently superior to the pendulum for recording very long period waves because of its lack of response to tilt and because the strain response varies inversely as the period, whereas the pendulum response varies inversely as the square of the period. The electromagnetic strain seismograph is equipped with a variable reluctance transducer essentially identical with that described for the pendulum seismograph. Figure 23 is a photograph of the transducer end of the Pasadena strain seismograph showing the pier and the end of the indicator tube. The transducer is provided with eight coils, as in the case of the pendulum instrument, and serves to operate two or more galvanometers simultaneously.

When provided with an electromagnetic transducer, the period-response characteristic of the strain seismograph is identical with that of the simple pendulum seismograph having a pendulum period and

FIG. 23. Photograph of transducer end of strain seismograph—Benioff.

damping constant equal to the period and damping constant of the galvanometer. Figure 24 shows the period-response characteristic of the electromagnetic strain seismograph for the two cases where the galvanometer damping is critical, $h = 1$, and for the case where $h = \frac{1}{2}\sqrt{2}$. The latter case corresponds to the value of damping giving the widest range of periods over which the response is flat. For most purposes, therefore, the value $h = \frac{1}{2}\sqrt{2}$ is to be preferred. These curves also represent the response characteristics of any pendulum seismograph in which the trace amplitude is proportional to the pendulum displacement. The remarks concerning damping are applicable in this case also. The period response of the electromagnetic strain seismograph depends upon the galvanometer only, and consequently a wide variety of response characteristics is available by merely changing galvanometers. Thus with a galvanometer period of 0.8 seconds the period-response characteristic is equivalent to that of the Anderson-Wood short-period torsion seismo-

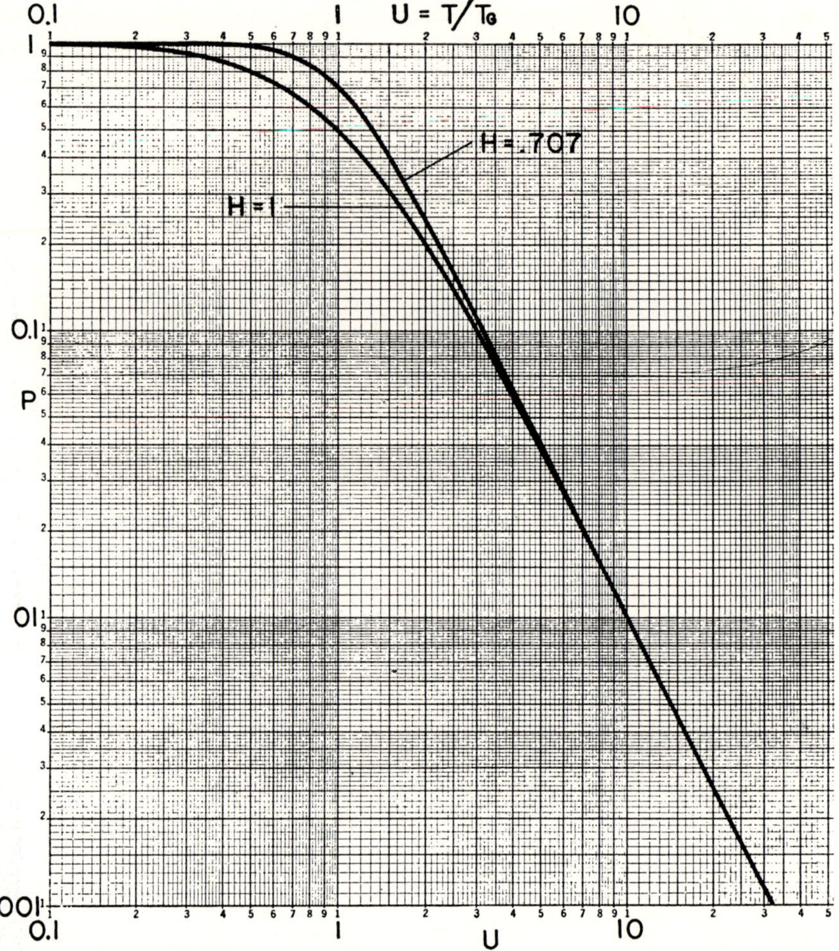

FIG. 24. Period-response characteristics P_1 and P_2 of an electromagnetic linear-strain seismograph for critical galvanometer damping ($H = 1$) and for the value of damping for which the characteristic is flat over the widest period range ($H = \frac{1}{2}\sqrt{2}$). These are also the response characteristics of a pendulum seismograph with optical or mechanical recording and with pendulum damping constants as indicated. T = period of ground vibration, T_g = period of galvanometer, $P_1 = \frac{1}{u^2 + 1}$ for $H = 1$, and $P_2 = \frac{1}{(u^4 - 1)^{1/2}}$ for $H = \frac{1}{2}\sqrt{2}$.

graph. Since the superiority of the strain seismograph over the pendulum seismograph increases with period, it has been found desirable at the Pasadena laboratory to operate the strain instrument routinely with galvanometer periods of 70 and 180 seconds respectively.

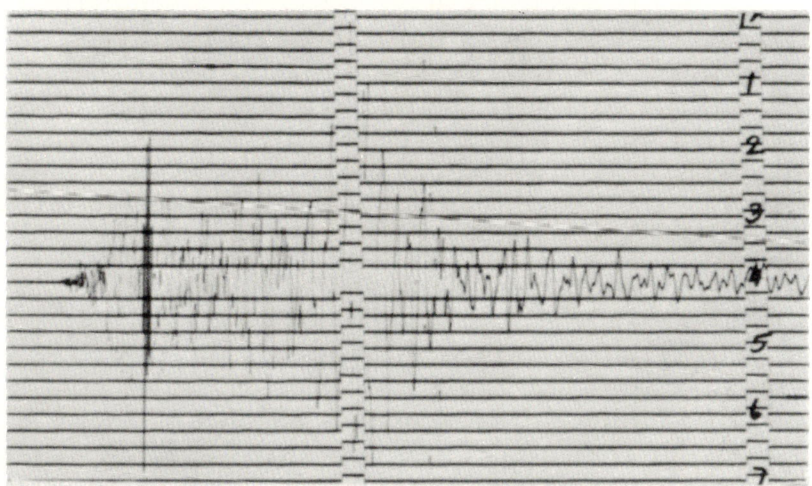

Fig. 25. Strain seismogram of a small local earthquake, $\Delta = 35$ km, recorded at Pasadena. $T_g = 0.25$, $H = 1$, and $V = 80{,}000$ (approximately).

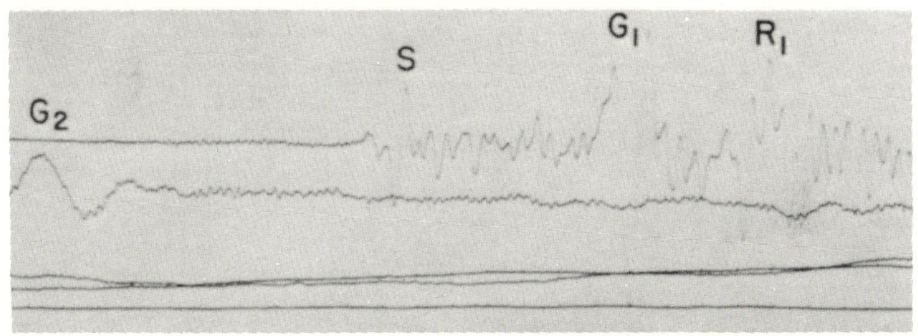

Fig. 26. Portion of strain seismogram of Kamchatka earthquake of November 4, 1952, recorded in Pasadena. $T_g = 180$ seconds, $V = 20$ (approximately). Note 7- to 8-minute waves riding on the lower displaced line, which is a portion of a wave of 57-minutes period.

Figure 25 shows a portion of a seismogram of a small local earthquake ($\Delta = 35$ km approximately) recorded at Pasadena with the short-period galvanometer strain seismograph having a galvanometer of 0.25-second period, critical damping, and an equivalent pendulum static magnification of approximately 80,000. Figure 26 shows a portion of a seismogram of the Kamchatka earthquake of November 4, 1952, written with the 180-second period galvanometer on the strain seismograph, showing impulse forms of S, G_1, R_1 and G_2. The displaced zero line appearing in this seismogram is actually a portion of a 57-minute period wave group and represents the longest period wave recorded to date with any seismograph. Since the length of the steel indicator tube varies with temperature to the extent of approximately 1 part in 10^5 per degree centigrade, it has been found desirable to surround the tube with asbestos or another thermal insulator to reduce temperature fluctuations in length. For short period galvanometer registration, the thermal variations are negligible. As the period of earth strains is increased, however, the thermal expansion becomes more and more important until for very long periods it represents the principal limitation to the useful magnification of the instrument.

7.2. Fused Quartz Strain Seismograph—Benioff

In order to push the response to longer periods and to record permanent strains in the ground such as those which give rise to earthquakes, a linear strain instrument was constructed using quartz tubing for the indicator [21]. It is housed in a tunnel bored in the mountains in Dalton Canyon northeast of Pasadena. Figure 27 is a schematic sectional drawing

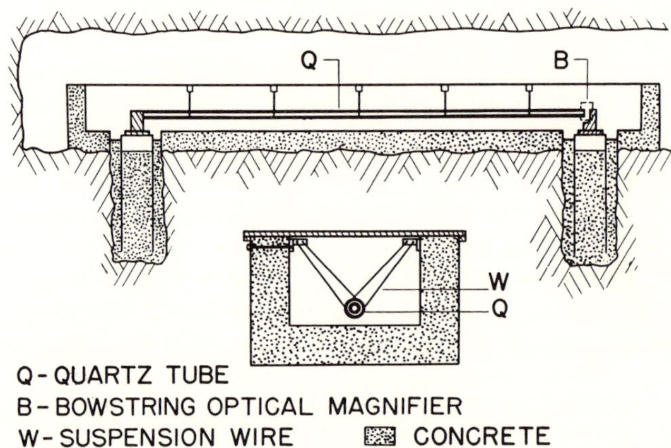

FIG. 27. Schematic sections of the Dalton fused quartz, secular strain seismometer.

of the instrument, showing the piers, which are made of 10-inch steel pipes cemented into the rock, and the supporting wires, constructed of stainless steel 0.012 inches in diameter. In the original installation the instrument is provided with a double bow-string optical magnifying system for photographic recording. Figure 28 shows a schematic drawing of the bow-string magnifier. The suspensions are made with stainless steel wire 0.001 inch in diameter and are approximately 20 centimeters long overall. Movement of the quartz tube relative to the pier tightens one suspension and loosens the other, thus rotating the mirror. If b is the half-length of the suspension, a is the vertical depression of the mirror

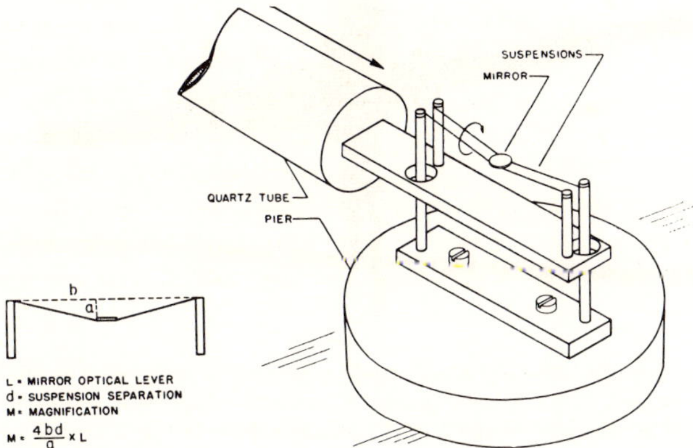

FIG. 28. Schematic representation of double bowstring optical magnifier for secular strain seismometer. The suspension system is damped electromagnetically by means of the field of a permanent magnet not shown in the drawing.

relative to the suspension supports and d is the distance between suspensions, the magnification is given by

$$V = \frac{4bdL}{a} \tag{7}$$

where L is the optical lever (the distance from the mirror lens to the recording drum). In the Dalton installation the constants were chosen to give a value of magnification V of 10,000. Since the length of the strain instrument is 24.08 meters, this value of magnification provides a trace amplitude of 1.0 mm for a strain of 4.15×10^{-9}. Figure 29 is a copy of a seismogram written with the Dalton quartz strain instrument of the Japanese earthquake of November 25, 1953. The recording speed was 1 cm per hour. Figure 30 is a recording made with the same instrument showing the tidal strains of the earth as produced by the sun and moon.

Since the response of the strain seismograph is of first order for local strains produced within a distance comparable to its length and is second order for seismic wave strains, it is subject to greater disturbances from the effects of persons moving about in the vicinity of the instrument than is the pendulum instrument. Consequently, a strain seismograph should be located preferably at some distance from occupied buildings, roads, and other sources of disturbing strains. The instruments which have been constructed to date have been mounted in weathered granite.

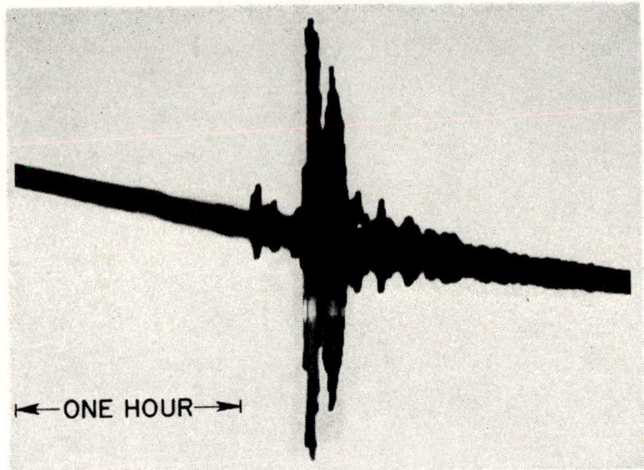

FIG. 29. Recording of the Japanese earthquake of November 25, 1953 $M = 8.25$, written by the Dalton fused quartz, secular strain seismograph. The recording speed was 1 cm/hour. The diagonal slope of the mean position of the trace is due to the tidal strain variation. Note the small but definite discontinuous change of slope during the passage of the earthquake wave-train, possibly indicating a worldwide strain adjustment in response to the strain release at the focus.

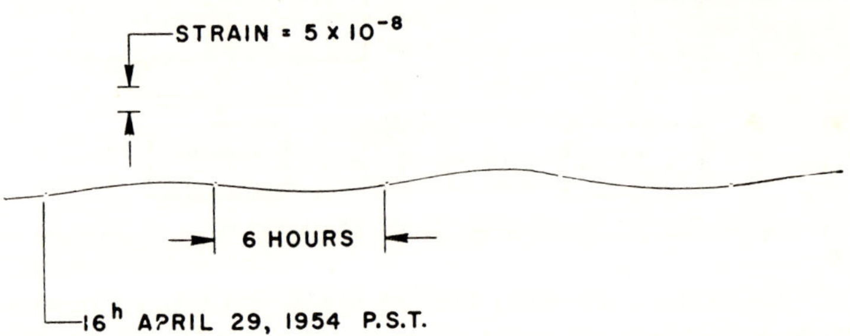

FIG. 30. Recording, made with the Dalton fused quartz, secular strain seismograph, showing tidal strains of the earth.

It is not known how well they perform if mounted on sedimentary rock or on soil.

8. Remote Recording Seismographs

8.1. Radio Telerecording Seismograph—Gane, Logie, and Stephen

A system for recording a number of seismometers situated up to distances of 17 kilometers from a base recording station has been described by P. G. Gane, H. J. Logie, and J. H. Stephen [22]. The purpose of this development was the recording of earth waves produced by rock bursts in the Witwatersrand mine. Consequently the recorded seismic frequencies were limited to the range from 5 to 20 cycles per second. In many respects this limitation to higher frequencies simplifies the recording problems. The principles developed are sound, however, and with slight modification can be used for recording frequencies encountered in natural earthquakes. The seismometers are electrostatic instruments developed by P. G. Gane [15]. After two stages of amplification the seismometer output serves to modulate the frequency of an 800-cycle multivibrator oscillator, the maximum available deviation being approximately ± 300 cycles for 1 percent distortion. The 800-cycle frequency-modulated multivibrator output serves in turn as a subcarrier to amplitude modulate a 41 to 46 megacycle radio transmitter of 15 watts power. The base station is provided with separate radio receivers tuned respectively to each of the field transmitters. The output of each receiver

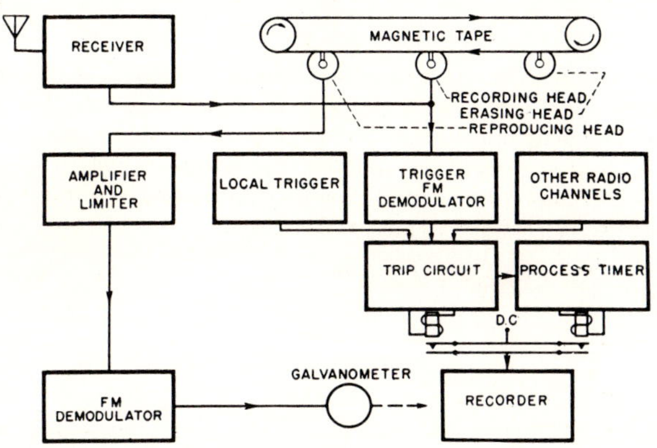

FIG. 31. Delayed telerecording seismograph system of Gane, Logie, and Stephen.

(Fig. 31) consists of the 800 cycle fm subcarrier signal. In order to be able to record at a high speed (7 mm per second) without the use of excessive amount of film, the authors introduced a memory system in the form of a 96 foot continuous loop of steel magnetic tape moving continuously at a speed of 30 inches per second. A portion of the 800 cycle fm subcarrier from the receiver is recorded continuously on the tape with a standard tape recording head. The rest, after demodulation, serves to actuate a trigger circuit normally inoperative until a signal corresponding to a seismic wave at one of the field stations is received. The trigger in turn actuates a multiple paper recorder, which has one galvanometer for each field station and is normally in standby condition. A reproducing head is positioned to the rear of the recording head at a sufficient distance to provide a 6.2 second delay between the recorded and the reproduced signals. After amplification and limitation the 800 cycle fm signal from the reproducing head is demodulated to seismic frequency and then serves to actuate the galvanometer. An erasing head precedes the recording head. The recorder thus operates only after a rockburst occurs and it continues to operate for one minute only unless restarted by a subsequent shock. The amplitude range of the equipment exclusive of the tape unit and recorder is approximately 800 to 1. Owing to variations in the speed of the magnetic tape, the recorded range is limited to about 200 to 1. In the assembly described by the authors, six galvanometers recorded simultaneously the movements of the six field seismometers. The galvanometer frequencies were 15 cycles per second. Operating without the tape delay unit, the maximum available magnification at the recorder was 15,000. With the memory unit in place this was reduced to 5,000 as a result of fm modulation of the recorded subcarrier by variation in speed of the magnetic tape. The precision in timing of the arrival of seismic waves (approximately 0.01 second) was higher in this network, by an order of magnitude, than that which has been available with other networks involving conventional recording and timing at each field station. If this system can be modified to operate up to distances of 400 or 500 kilometers and with period ranges extending to ten seconds or longer, it will represent a great advance over existing local earthquake networks. In addition to the increased timing accuracy, the recording of all seismometers of a network at a single base station on a single strip of paper or film provides conveniences in measurements, station to station comparisons, and supervision which cannot be overestimated.

It should be pointed out that the magnetic tape memory unit with delayed recording can be operated directly by a seismometer without the radio link. As such it provides maximum recording speed with no waste of recording film or paper. However, it is effective only for the recording of shocks having maximum amplitudes substantially larger than the microseism amplitudes.

9. Components

9.1. Wilmore Pendulum Suspension

Wilmore [23] developed a short-period moving-conductor transducer seismograph having a novel linear displacement pendulum suspension. In older forms of linear motion pendulums, such as the variable reluctance seismographs designed by the writer, the suspension consists of six taut ribbons or wires extending from the pendulum outward to fixed supports (see Fig. 6). In order to keep the restoring force introduced by these ribbons to reasonable values, they must be fairly long and thus the overall dimensions of the instrument are substantially increased. Wilmore's suspension occupies very little more space than that of the pendulum itself (see Fig. 32). It consists of six members, mounted in two groups of three at each end of the pendulum. The members consist of short light rods R attached to frame lugs A, B, and C and to pendulum lugs a, b, and c by means of short lengths of wire W, soldered into the rods and into slots cut in the lugs. These six members thus constrain the inertia reactor to move linearly parallel to its axis. The wires are just large enough in diameter to withstand buckling, and consequently they introduce only a relatively small amount of restoring force into the system. Since the pendulum movement is linear, the restoring force of the suspensions is independent of orientation of the pendulum, and

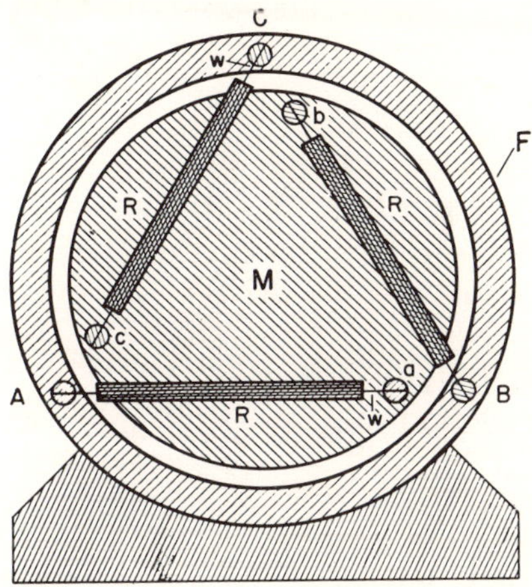

FIG. 32. Wilmore seismometer pendulum suspension (schematic).

consequently the same structure is suitable for both vertical and horizontal component pendulums. The linear motion of the pendulum is accompanied by a slight rotation, which in most applications is of negligible importance.

9.2. LaCoste Vertical-Component Pendulum

A significant improvement in the deflection stability of long-period vertical-component pendulums was made by Lucien LaCoste [24]. For vertical pendulums of the linear type in which the inertia reactor is suspended directly by the spring, the spring extension is given by

$$Z_0 = \frac{gT_0^2}{4\pi^2} \tag{8}$$

where g is the acceleration of gravity and T_0 is the pendulum period. Thus for a pendulum period of 10 seconds the spring extension is approximately 25 meters. Since it is impractical to construct pendulums with such large spring extensions, means have been found for obtaining long periods by introducing negative restoring force to compensate the positive restoring force of springs of relatively short length. These methods, however, are not entirely satisfactory in that the negative restoring force is nonlinear in relation to the pendulum deflection. Thus, for deflections of the pendulum greater than a certain small value, the negative restoring force becomes greater than the positive, and the pendulum is rendered unstable. The LaCoste pendulum is constructed with an inertia reactor W (Fig. 33) mounted on a boom pivoted at D. The spring is attached to the frame at F, a point vertically above D, and to the boom of the pendulum at the point C. The innovation introduced by LaCoste was

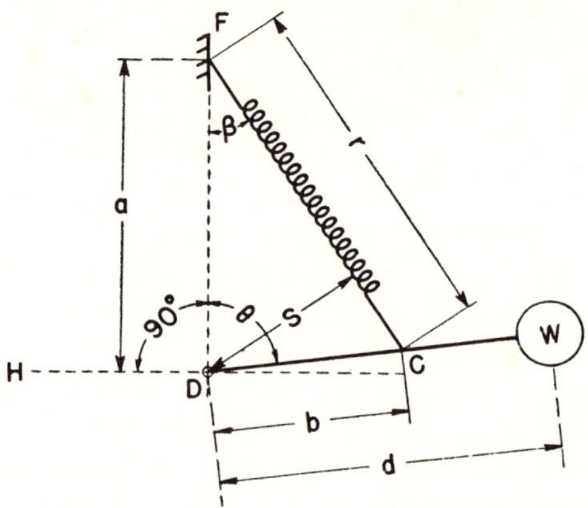

FIG. 33. LaCoste pendulum.

the use of a zero-length spring—a spring in which the actual physical length is equal to the elongation. When unloaded, such a spring has a substantial residual tension. A special procedure for winding the spring is described by LaCoste. Referring to the figure, the tension of the spring is kr, where k is the spring constant. The proof for the condition of stability with infinite period as given by LaCoste is as follows: The torque produced by the weight is

$$N_w = Wd \sin \theta \tag{9}$$

Since the spring has a zero initial length, the torque produced by the spring is

$$N_s = -krS \tag{10}$$

where S is the perpendicular distance of the spring from the axis of rotation. Furthermore

$$r = \frac{b \sin \theta}{\sin \beta} \tag{10.1}$$

and

$$s = a \sin \beta \tag{10.2}$$

The expression for N_s becomes

$$N_s = -kab \sin \theta \tag{10.3}$$

The total torque is thus

$$N = N_w + N_s = (Wd - kab) \sin \theta \tag{11}$$

If

$$Wd = kab \tag{12}$$

then

$$N = 0$$

for all values of θ, and the period is infinite for all positions of the boom. Thus, theoretically speaking, the instrument is stable for angular deviations of the boom of 90 degrees above and 90 degrees below the horizontal. Actually, of course, the pendulum does not behave ideally because of pivot friction, the restoring force added at the points F, C, and D, and the distortion of the spring by gravity. Nevertheless LaCoste found that with a period of 37 seconds, the period was constant to within one second for boom deflections of 9 degrees—a great improvement over previous systems. It should be pointed out, however, that with the best of alloy springs the LaCoste pendulum, like other vertical pendulums, is responsive to temperature variations and to creep. However, for most alloys the creep rate decreases with time.

9.3. Factors Affecting Performance of Photographic Recorders

Most present day seismographs record on photographic paper using the crossed cylindrical lens optical system described in connection with the Anderson-Wood torsion seismograph and shown in Fig. 1. In describing the performance of a seismograph, a mere statement of the magnification is meaningless unless the size and sharpness of the recording spot is known. The available information, or resolution, of a seismogram thus depends upon the ratio of magnification to the recording spot size and sharpness, as well as upon the ratio of recording speed to earth period. A desirable shape for the recording light spot is a rectangle having a length approximately 5 to 10 times the width, oriented with the width parallel to the time direction. With this shape, maximum precision is maintained in time measurements. Furthermore, during rapid excursions of the light spot, the recorded intensity is maintained at a higher level than can be achieved with a circular or square spot. If the time of onset of a phase is to be measured to 0.1 second, say, then the width of the light spot rectangle in the time direction must be less than the distance the recording spot travels in 0.1 second in the quiescent state. Thus, for recorders operating at a speed of 1 mm per second, the time dimension of the light spot should be less than 0.1 mm for 0.1 second accuracy. On the other hand, with a film recorder moving at the rate of 0.25 mm per second, the recording light spot should have a time dimension of less than 0.025 mm (0.001 inch). Another factor affecting the precision of time measurements on seismograms, which is frequently overlooked, is the accuracy of the gears which drive the recording drum. For a drum with a circumference of 900 mm and a linear speed of 1 mm per second, or one revolution in 15 minutes, a precision of 0.1 second of time requires an accuracy in the final gear of approximately 6×10^{-4} radian, and this is difficult to achieve without special machine work.

9.4. Tuning Fork Drive for Recorders—Benioff

In earlier days when the powerline frequency was not controlled, and at the present time in places where controlled power is not available, accurate rotation of the drum was and is kept by means of tuning fork controlled synchronous motors. One such system which has been in service for nearly thirty years at the Seismological Laboratory in Pasadena [25] consists of a tube-maintained tuning fork, whose frequency is 10 cycles, for controlling the operation of two polar relays in quarter phase relation. The polar relays in turn supply forty accurately timed impulses to special impulse-type synchronous motors. The synchronous motors (Fig. 34) have four soft iron four-pole rotors mounted on a common shaft, with angular displacements of 22.5 degrees between successive rotors. Each rotor is positioned in the field of a U-shaped stator. Impulses from the relays to the stator windings thus provide synchronous operation of the motors at a speed of 2½ revolutions per second. The power required per motor is approximately ⅓ watt and is derived from a 12-volt storage battery continually charged by a rectifier from the powerline. The fork is enclosed in a thermally insulated box and maintained at a constant temperature to within $\pm 0.1°$ C. The speed of rotation is approximately constant to within one part in 100,000. In normal operation, one set of relays operates up to four recording drums. The relays usually operate over periods of six months to one year without adjustment or dressing of the points. A circuit diagram of the fork system is shown in Fig. 35. In the figure, L_1 and L_2 are plate and grid coils which surround, without touching, two cylindrical permanent magnets screwed to the tines of the fork. C_5 and L_3 form a 10-cycle resonant coupling circuit from the vacuum tube VT_2. The resistances and condensers R_7, C_7, R_8, C_8 represent a phasing network for providing a 90 degree phase lead of the grid voltage of VT_3 relative to that of VT_2. The polar relay SW_1 is driven by VT_3, and the relay SW_2 is driven by

FIG. 34. 10-cycle synchronous impulse motor with gear train—Benioff. Motor speed is 2.5 rps, final shaft speed 1 rpm.

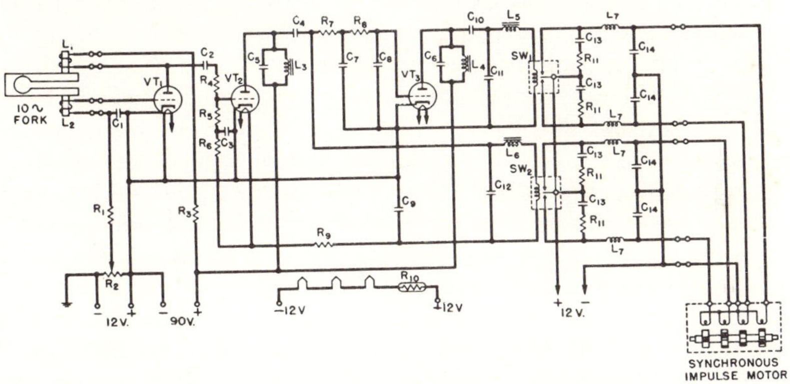

FIG. 35. Circuit diagram 10-cycle fork-controlled recording drum drive—Benioff.

VT_2. The condensers and resistors C_{13} and R_{11} serve to reduce arcing of the relay points. L_7 and C_{14} are interference filters for reducing radiation disturbance to nearby radio receivers. A ballast lamp R_{10} maintains the current to the heaters of the tubes constant. The anode voltage supply for the tubes may be either a dry battery or a rectifier operated from the powerline and provided with a means to change over to a standby battery in case of line failure.

9.5. Photographic Paper Recorders

Figure 36 is a photograph of the paper recorder of the short-period galvanometer combination of the variable reluctance seismograph which

FIG. 36. Short-period galvanometer recorder—Benioff. $T_g = 0.2$ sec, drum speed-1 mm/sec.

has been in use for many years at the Seismological Laboratory in Pasadena. The galvanometer has a period of 0.23 seconds and was designed by William Miller, formerly of the laboratory staff. The light source is near the base of the drum, and the optical system is so arranged that the light spot falls on the top surface of the drum in order to provide ease of inspection and adjustment. Figure 37 is a photograph of a recording drum of recent design manufactured by the Geotechnical Corporation in cooperation with the writer. This drum is arranged to be entirely enclosed, if necessary, in order to operate in a lighted room. Figure 38 shows the paper recorder of the long-period combination of the vertical-component variable reluctance seismograph in operation at the Pasadena Seismological Laboratory. The galvanometer, designed by William Miller, has a period of 115 seconds. The base of this galvanometer is

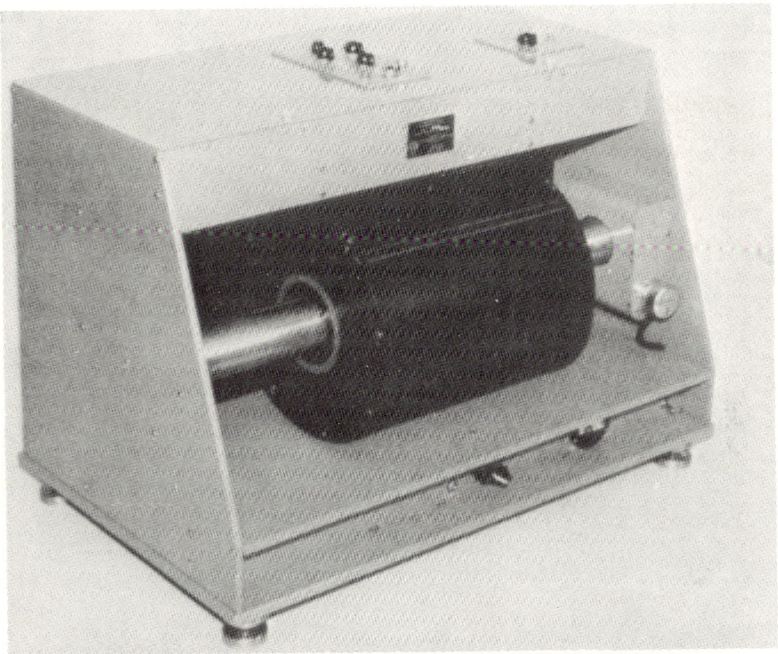

FIG. 37. Seismograph recorder—Geotechnical Corporation.

arranged with a bearing and worm screw which permits rotation of the galvanometer as a whole for fine adjustment of the light spot. A torsion head on the suspension is provided for adjustments which go beyond the range of the worm.

9.6. Film Recorders

As a result of the high cost of photographic paper and the problems connected with storage of large numbers of seismograms, many recorders in recent years are being operated on photographic film. The first of such recorders developed by the writer is shown in Figure 39. The film is a 91-centimeter length of 35 millimeter positive motion picture film. It is wrapped around a narrow recording drum and clamped with approximately 1 centimeter of overlap. The effective light source is a real image of an automotive-type tungsten lamp filament, reduced 8 to 1 by a microscope objective. The recording speed is 0.25 mm per second, and the advance of the drum is 1 mm per revolution. A single film can thus run for about 27 hours. A crossed cylindrical lens system is used, with foci reduced approximately 4 to 1 in relation to the paper recorders. In spite of the slower recording speed and smaller size of the 3.5 × 90 cm

FIG. 38. Long-period (115 seconds) galvanometer recorder of the Seismological Laboratory.

film strip, as compared to the 30 × 90 cm sheet in photographic paper recorders, the higher resolution of film results in an actual increase in the amount of recorded information. A late model recorder manufactured by the Geotechnical Corporation of Dallas, Texas, is shown in Fig. 40. This recorder operates four drums on the single shaft and is provided with all necessary controls for recording lamp intensity, time mark amplitude, galvanometer sensitivity, and galvanometer damping.

Fig. 39. Original seismograph film recorder—Benioff.

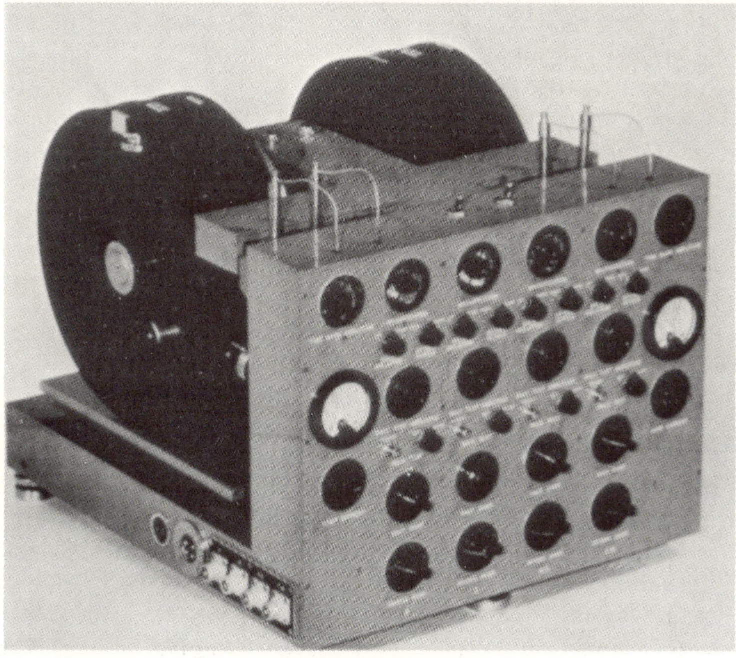

Fig. 40. Four-channel seismograph film recorder—Geotechnical Corporation.

9.7. Ink Writing Recorders

A number of visible writing recorders have been developed in recent years as supplementary instruments or substitutions for the photographic recorders in general use. In resolution and convenience the visible recordings are very nearly equivalent to the photographic paper recordings. For many years the Seismological Laboratory of the California Institute of Technology operated an ink writer representing a modification of the one described by Halley Wolf [26]. In this instrument the light beam from a 14 second period galvanometer connected to a vertical-component variable reluctance seismograph falls on two adjacent prisms mounted in such a way as to direct equal portions of the beam to two photocells when the galvanometer is in the quiescent position. The photocells are connected to a push-pull two-stage resistance-capacitance coupled amplifier having a time constant of approximately 20 seconds. Deflections of the galvanometer alter the light intensities on the two photocells in opposite phase and thus produce changes in the potentials across the photocell coupling resistors. If the galvanometer movements occur at seismic frequencies, the photocell outputs are amplified by the amplifier. The push-pull output of the amplifier is connected to the two halves of a center-tapped coil of a large writing galvanometer designed by William Miller. The writing galvanometer is critically damped by the plate resistances of the output tubes. The coil of this galvanometer is approximately 3 inches square and is suspended by a heavy bronze ribbon. The writing galvanometer period is 0.2 seconds approximately. The pen which is attached to the upper part of the coil assembly is made entirely of glass, and consists of a cylindrical reservoir for ink, approximately one centimeter in diameter by 3 centimeters in length. The cylinder is drawn down in a gentle taper to a fine point bent at right angles to the length of the tube. The total length of the pen is approximately 12 inches. With the development of the variable discriminator seismograph, the visible recorder at the Pasadena Seismological Laboratory was revised to operate with this instrument in accordance with the circuit diagram shown in Fig. 19. The original Miller writing galvanometer was retained.

9.8. Hot Stylus Recorders

Another form of visible writer, developed by the Sanborn Company for use with electrocardiographs, uses a heated stylus in combination with a recording paper having a black base with a white coating, presumably plastic. Contact of the heated stylus on the paper melts the white coating at the point of contact, and thereby exposes the black background. The recording galvanometer has a period of approximately 0.03 second, a coil resistance of 3000 ohms and a sensitivity of approximately 2.5 centimeters deflection for a current of 30 milliamperes. The paper comes in rolls 200 feet in length and moves continuously through the recorder. In the original form the instrument is not suitable for routine seismographic recording because of the great amount of paper required. For portable service, such as recording of aftershocks near their epicenter, or for short runs, it is entirely satisfactory and can be

FIG. 41. Hot stylus seismograph paper recorder—Geotechnical Corporation.

operated with the variable discriminator circuit shown in Fig. 19. For this service the recording speed is either 1 or 2 mm per second. A modified form of this recorder, in which a single 30 × 90 cm sheet is wound about a standard drum, was developed by the Geotechnical Corporation and is shown in the photograph in Fig. 41. In this type of recorder it is not feasible to overlap the ends of the paper, and consequently a mechanism was designed which permits the two ends to pass through a narrow slit in the drum and to engage with members within the cylinder for tightening the sheet.

9.9. Film Projection Reader—Benioff

Since the seismogram written on a film recorder is too small for direct measurement, two types of measuring devices have been designed employing optical magnification. In one an 8-fold enlarged image of a portion of the film is projected onto a ground glass for observations and measurement with manual scales. Figure 42 is a schematic representation of the projector. Light from an incandescent filament passes through a condenser and is reflected vertically downward to the film, which lies flat horizontally. The 72-mm focal length lens, mounted below the film, projects an image on the ground glass after reflection from the large mirror situated below the lens. For film seismograms recording at the standard rate of 0.25 mm per second, the projected image has a time scale of 2 mm per second of time. During operation the film is compressed

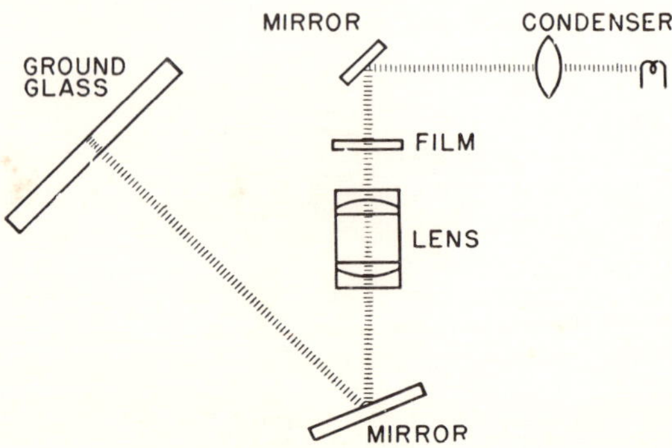

FIG. 42. Projector for measuring 35 mm film seismograms (schematic)—Benioff.

between two glass plates the pressure of which is controlled by a foot pedal. Pressure on the foot pedal releases the film and thus allows manual positioning to the region of study. The ground glass is removable and can be replaced with a film holder for making enlarged copies of portions of the seismogram.

9.10. Film Measuring Microscope—Benioff

For more precise measurements of time and amplitude on film seismograms, the measuring-microscope assembly shown in Figs. 43 and 44 has been designed. In use, the entire film is clamped along its edge over a rectangular strip of plate glass and is illuminated from below by a fluorescent lamp, which extends the full length of the film. The binocular microscope is mounted on a carriage which can be moved along the length of the film and clamped at any position of interest. The carriage is provided with a double-slide micrometer assembly, which reads directly in seconds and tenths of seconds in the time direction, and in millimeters and tenths in the amplitude direction. Figure 43 is a photograph of the complete assembly, and Fig. 44 shows the microscope and double slide carriage.

9.11. Quartz Crystal Clock—Benioff

For many seismological investigations, an accurate clock is as important as the seismograph itself. In regions which are not subject to strong earthquakes, good pendulum clocks make satisfactory time-keeping elements. In seismically active regions such as California, it has been the practice in the past to use marine type chronometers. These are fairly satisfactory if two or more radio time signals are available daily. These time pieces require servicing approximately once per year, and in recent years this has been rather costly. In order to improve on the

Fig. 43. Measuring-microscope assembly for 35 mm film seismograms—Benioff.

marine chronometer, a quartz crystal clock has been recently developed at the Seismological Laboratory of the California Institute of Technology [21]. The timing element of the clock is a 100-kilocycle quartz standard oscillator manufactured by the Bliley Company (Fig. 45). This unit has a low temperature-coefficient crystal housed in a constant temperature oven. The counter, manufactured by the Wang Company, is made up of a series of saturating transformer units known as "Perma-Memory Multiple Scalers." It delivers one output impulse for each six million oscillations of the crystal, corresponding to one impulse per minute. The duration of the impulse is approximately 5 microseconds, and this is lengthened to 1 second by means of a monostable multivibrator circuit called a pulse-stretcher. The lengthened impulse from the pulse-stretcher serves to operate the seismograph time-marking relays and in addition actuates an I.B.M. impulse clock for providing a visible indication of the time. The I.B.M. clock is arranged with a cam mechanism for making contacts once per hour which actuate the time-marking relays for hour

FIG. 44. Measuring microscope for 35 mm film seismograms showing double-slide carriage with calibrated time- and amplitude-screws.

marks. In order to insure uniform action and long life of the vacuum tubes, a constant voltage transformer is used to reduce powerline voltage fluctuations. According to the manufacturer's specifications, the oscillator is accurate to within one part in 10^7 in 24 hours—better than 0.1 second per day as a clock. In running tests the clock keeps time well within 0.1 second per week.

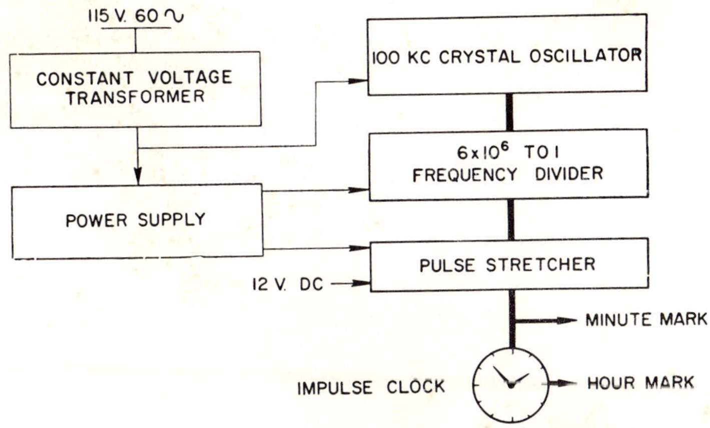

FIG. 45. Block diagram of 100-kc quartz crystal clock—Benioff.

10. SEISMOGRAPH RESPONSE CHARACTERISTICS

10.1. Response of Vertical-Component Seismograph to Barometric Pressure Variations

A. P. Crary and Maurice Ewing [27] observed that a disturbance, having a period of approximately 5 minutes, on a seismogram written by a vertical-component seismograph had occurred simultaneously with a similar disturbance recorded on a microbarograph. They subsequently developed the theory of the response of a vertical pendulum to barometric pressure variations and found that the predicted response of the pendulum agreed with the observations. Changes in the barometric pressure are accompanied by changes in density of the air and this in turn produces changes in buoyancy on the pendulum. According to Crary and Ewing, if the pressure P has the form

$$P = P_0 + p \sin \omega t \tag{13}$$

where P_0 is the quiescent barometric pressure in millibars (1019 mb) and p is the amplitude of the pressure variation in millibars, then the acceleration of a pendulum due to the varying buoyancy is

$$\ddot{Z}_a = \frac{\rho_0 g}{\rho_m P_0} p \sin \omega t \tag{14}$$

In this equation ρ_0 is the quiescent atmospheric density (0.00136 gr/cm^3), ρ_m is the density of the inertia mass (7.8 gr/cm^3 for iron), and g is the acceleration of gravity. Setting

$$\frac{\rho_0 g}{\rho_m P_0} = b \tag{14.1}$$

Equation (14) becomes

$$(15) \qquad \ddot{Z}_a = bp \sin \omega t$$

If the pendulum is disturbed by a seismic wave having the same period as that of the pressure variation in which the ground displacement is given by

$$(15.1) \qquad \xi = \xi_0 \sin \omega t$$

The resulting acceleration of the pendulum is

$$(16) \qquad \ddot{Z}_s = -\xi_0 \omega^2 \sin \omega t$$

By equating the accelerations of equations (15) and (16) we may determine the value of the ground displacement which would produce the same disturbance on the seismograph record as the given barometric variation. Thus neglecting the negative sign of (16)

$$(16.1) \qquad bp \sin \omega t = \xi_0 \omega^2 \sin \omega t$$

and the ground displacement amplitude corresponding to a pressure variation amplitude p is

$$(17) \qquad \xi_0 = \frac{bp}{\omega^2} = \frac{bpT_p{}^2}{4\pi^2}$$

where T_p is the period in seconds of the pressure variation, as well as that of the equivalent ground displacement. With an iron pendulum $b = 1.68 \times 10^{-4}$, and the expression for the equivalent ground displacement is

$$(18) \qquad \xi_0 = 4.56 \times 10^{-6} \times pT_p{}^2$$

Equation (17) shows that the equivalent ground displacement for a given pressure variation varies with the square of the period, and that the disturbance produced by pressure variations is more pronounced with long period seismographs that it is for short period instruments.

The pressure variations of 5-minute periods observed by Crary and Ewing had an amplitude of approximately 1 millibar. The equivalent ground motion was thus

$$(19) \qquad \xi_0 = 4.56 \times 10^{-6} \times 9 \times 10^4 = 4.1 \times 10^{-1} \text{ cm}$$

Since the sensitivity of most seismographs to waves of 5-minute period is very low, the response is not large even though the equivalent displacement is. For example, applying the results of equation (19) to the long-period galvanometer combination of the variable reluctance seismograph, we find from curve VR_2, Fig. 46, that the magnification for waves

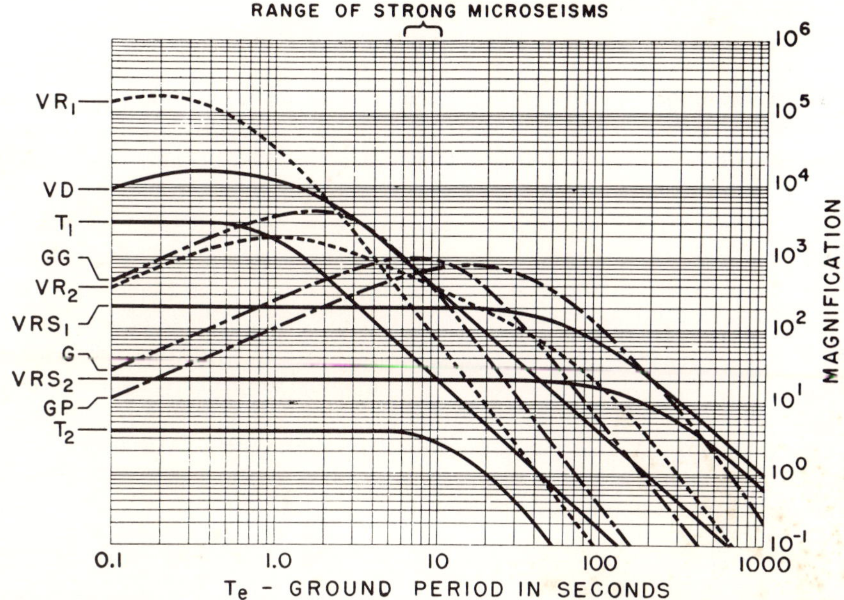

FIG. 46. Typical period-response characteristics of representative seismograms in terms of ground-particle displacement. T_0 is the pendulum period in seconds; h_0 is the damping coefficient of the pendulum (for critical damping, $h = 1$); T_g is the galvanometer period in seconds and h_g is the galvanometer damping coefficient.

VR_1—Benioff moving conductor electromagnetic seismograph. $T_0 = 1$ second, $h_0 = 1$, $T_g = 0.2$, $h_g = 1$. This curve also represents the Benioff variable reluctance seismograph having the same pendulum and galvanometer periods and damping coefficients. For this latter combination the curve is not accurate for periods less than 1 second. See Fig. 7 for the measured response in this period range.

VD—Benioff variable discriminator seismograph. $T_0 = 1.5$, $h_0 = 1$, $T_g = 0.1$, $h_g = 1$.

T_1—Anderson-Wood short-period torsion seismograph. $T_0 = 0.8$, $h = 1$, $V = 2700$.

GG—Gutenberg modification of the Galitzin seismograph. $T_0 = T_g = 3$ seconds, $h_0 = h_g = 1$.

VR_2—Benioff variable reluctance seismograph. $T_0 = 1$, $h_0 = 1$, $T_g = 90$, $h_g = 1$.

VRS_1—Benioff electromagnetic linear strain seismograph. $T_g = 70$ seconds, $h_g = 1$. It should be noted that the magnification of the strain seismograph depends upon apparent surface velocity of the wave. A given curve is thus strictly valid only for waves of a particular velocity.

G—Galitzin seismograph. $T_0 = 12.5$ seconds, $h = 1$. VRS_2—Benioff electromagnetic linear strain seismograph. $T_g = 180$ seconds, $h_g = 1$.

GP—Ewing-Press long-period moving-conductor electromagnetic seismograph. $T_0 = 15$ seconds, $h_0 = 1$, $T_g = 75$ seconds, $h_g = 1$.

T_2—Smith strong-motion torsion seismograph. $T_0 = 10$ seconds, $h_0 = 1$, $V = 4$.

of 5-minute period is unity. Hence the trace amplitude resulting from barometric variations of 1 millibar of 5-minute period is 4.1 mm.

In order to eliminate the response to pressure variation, Press and Ewing designed a vertical-component seismograph using a LaCoste pendulum having a hollow member. This member was positioned on an extension of the boom at a point opposite the steady mass with respect to the axis of rotation. The volume of the hollow member and its distance from the axis of rotation was chosen to balance the torque of the buoyant force on the rest of the pendulum. For pendulums having linear motion this type of compensation is not possible. These must be enclosed in a pressure-tight container. Such a container in the form of a cylindrical steel box was constructed at the Seismological Laboratory of the California Institute of Technology. In addition to preventing barometric responses, the pressure-tight container insulates the pendulum effectively from air currents and rapid temperature variations. Moreover, it can be provided with a drying agent such as silica gel for maintaining the air surrounding the pendulum in a dry state at all times, regardless of outside moisture conditions.

10.2. Response of Strain Seismometer to Barometric Pressure Variations

Under conditions of maximum sensitivity, the strain seismometer may show a measurable response to barometric pressure variations. Assuming that the tube is not closed, the linear strain of the tube in response to a pressure variation of p dynes/cm^2 is

$$\epsilon = \frac{p}{3k} \tag{20}$$

where k is the bulk modulus. If the length of the tube is L, the linear extension of the tube is

$$\Delta L = \epsilon L = \frac{Lp}{3k} \tag{21}$$

and the trace amplitude is

$$y = V\Delta L = \frac{VLp}{3k} \text{ cm} \tag{22}$$

where V is the magnification of the recording system. The bulk modulus k for the translucent fused quartz of the Dalton instrument is approximately 4.7×10^{11}. For iron, k is 1.7×10^{12} dynes/cm^2. The response of the Dalton strain instrument to a pressure variation of 1 millibar is thus

$$y = \frac{10^4 \times 2.4 \times 10^2 \times 10^3}{3 \times 4.7 \times 10^{11}} = 1.7 \times 10^{-3} \text{ cm}$$

10.3. *Typical Period-Magnification Characteristics of Representative Seismographs*

Calculated period-magnification characteristics of a number of representative seismographs are shown in Fig. 46. For many regions of the earth, the ground unrest in the period-range from 5 to 10 seconds is much

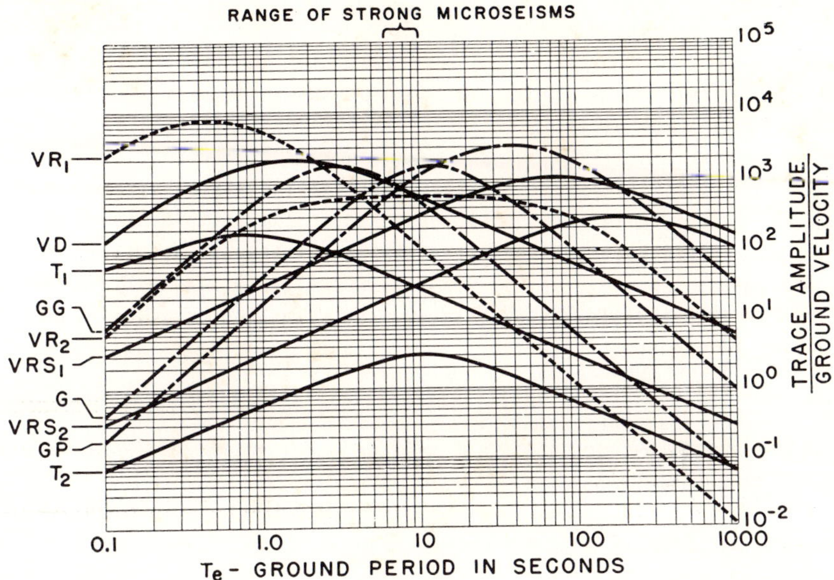

FIG. 47. Typical period-response characteristics of representative seismographs, in terms of ground-particle velocity. Instrument designations and constants are the same as in Fig. 46.

higher than it is for periods outside of this range. Consequently the maximum usable magnifications of seismographs are generally limited by the level of microseismic activity in this range. Throughout most of the year the microseismic level in Pasadena is such that magnifications in the critical range cannot exceed approximately 1000. In calculating the curves of Figs. 46 and 47 the constants have therefore been chosen with this value as the limit. For other localities, usable magnifications may be higher or lower depending upon the level of the ground unrest.

10.4. *Typical Ground-Particle Velocity-Response Characteristics of Representative Seismographs*

Figure 47 is a graph of calculated typical response characteristics, in terms of ground-particle velocity, for the same instruments as those represented in Fig. 46. It is the writer's contention that these characteristics are of greater significance than those of Fig. 46 expressed in terms of ground displacement. The particle velocity in a wave is of greater fundamental importance than the particle displacement, since the particle velocity is directly related to the power and energy of the wave. A seismograph having a flat characteristic on this diagram responds with the same trace amplitude to all waves of a given power regardless of their periods. Characteristics of this type are standard in acoustics.

List of Symbols

a distance
b distance
C horizontal wave velocity
d diameter
d distance
g acceleration of gravity
h damping factor
k spring constant or bulk modulus
l distance
L optical lever path, length of indicator, or length of tube
N total torque
N_s torque due to spring
N_w torque due to weight
p amplitude of pressure variation
P_0 quiescent atmospheric pressure
r length of suspension
S distance
t time
T ground period
T_0 characteristic period
T_g galvanometer period
T_P period of pressure variation
V magnification
W weight
y trace amplitude
Y_L response to longitudinal wave
Y_T response to transverse wave
Z_0 spring extension
Z_a acceleration due to buoyancy
$\ddot{Z}_s$ acceleration due to seismic disturbance
α angle
β angle
Δ distance of epicenter
ϵ linear strain
ξ displacement
ξ_0 ground displacement
ρ_0 quiescent atmosphere density
ρ_m density of inertia mass

References

1. Anderson, J. A., and Wood, H. O. (1925). Description and theory of the torsion seismolograph. *Bull. Seismol. Soc. Am.* **15,** 1–72.
2. Benioff, H. (1933). *Carnegie Inst. Year Book* **32,** 370.
3. Smith, S. (1930). *Carnegie Inst. Year Book* **29,** 436.
4. Wenner, F. (1932). Development of seismological instruments at the Bureau of Standards. *Bull. Seismol. Soc. Am.* **22,** 60.
5. McComb, H. E. (1936). Earthquake investigation in California 1934–1935. U. S. Coast and Geodetic Survey Special Publ. No. 201, p. 6.
6. Romberg, A. (1919). Theory of non-tilt seismograph. *Bull. Seismol. Soc. Am.* **9,** 135.
7. McComb, H. E. (1931). A tilt-compensation seismometer. *Bull. Seismol. Soc. Am.* **21,** 25.
8. Gutenberg, B. (1929). Sind Galitzin-Pendel für Nahbebenaufzeichnungen verwendbar. *Gerl. Beitr. Geophys.* **22,** No. 1 and 2, 100.
9. Wenner, F. (1929). *J. Research Natl. Bur. Standards* **2,** Research Paper No. 66.
10. Sprengnether, W. F. (1947). Horizontal and vertical component seismographs. *Bull Seismol. Soc. Am.* **37,** 101.
11. Ewing, M., and Press, F. (1952). Further study of atmospheric pressure fluctuations recorded on seismographs. Technical Report on Seismology No. 20, p. 7. (Lamont Geological Laboratory, Columbia University.)
12. Benioff, H. (1932). A new vertical seismograph. *Bull. Seismol. Soc. Am.* **22,** 155; see also *Carnegie Inst. Year Book* (1930) **29,** 434.
13. Benioff, H. (1933). *Proc. Pacific Sci. Congr. Pacific Sci. Assoc. 5th Congr. 1933* **3,** 2443.
14. Benioff, H. (1930). *Carnegie Inst. Year Book* **29,** 434, 435.
15. Gane, P. G. (1948). An electrostatic seismometer. *Bull. Seismol. Soc. Am.* **38,** 95.
16. Volk, J., and Robertson, F. (1950). The electronic seismograph. *Bull. Seismol. Soc. Am.* **40,** 81.
17. Cook, G. W. (1951). A sensitive vertical displacement seismometer. Report No. 814, The David W. Taylor Model Basin (U. S. Navy Department).
18. Benioff, H. (1951). Progress report, Seismological Laboratory, California Institute of Technology, 1950. *Trans. Am. Geophys. Un.* **32,** 749.
19. Sulzer, P. (1953). Transistor frequency standard, *Electronics* **26,** 206.
20. Benioff, H. (1935). A linear strain seismograph. *Bull. Seismol. Soc. Am.* **25,** 283.
21. Benioff, H. (1954). Progress Report, Seismological Laboratory, California Institute of Technology, 1953. *Trans. Am. Geophys. Un.* in press.
22. Gane, P. G., Logie, H. J., and Stephen, J. H. (1949). Triggered teleseismic recording equipment. *Bull. Seismol. Soc. Am.* **39,** 117.
23. Wilmore, P. G. (1950). The theory and design of two types of portable seismograph. *Geophysics* Suppl. **6,** 129.
24. LaCoste, L. (1935). A simplification in the conditions for the zero-length-spring seismograph. *Bull. Seismol. Soc. Am.* **25,** 176.
25. Benioff, H. (1926). *Carnegie Inst. Year Book* **25,** 431.
26. Wolfe, H. (1934). A seismographic recorder. *Rev. Sci. Instr.* **5,** 359.
27. Crary, A. P., and Ewing, M. (1952). On a barometric disturbance recorded on a vertical long period seismograph. *Trans. Am. Geophys. Un.* **33,** 499.

Author Index

Numbers in parentheses are reference numbers and are included to assist in locating references in which the authors' names are not mentioned in the text. Numbers in italics indicate the page on which the reference is listed.

C

D

E

F

G

Subject Index

AFFILIATED EAST WEST PRESS
(New Delhi)